Cheap Wage Labour
Race and Gender in the Fisheries of British Columbia

Using the fishing industry in British Columbia as a case study, Alicja Muszynski explores how Marx's labour theory of value can be applied to a specific industry and the creation of a specific labour force. She reworks Marx's theory in order to incorporate race and gender as principles that not only created a proletarianized labour force but also legitimized the payment of low wages to particular groups.

Cheap Wage Labour is the first analysis of shore work and shoreworkers in British Columbia from the 1860s to the mid-1980s. Muszynski provides an interpretation of the events that led to the creation of a cheap wage labour force of shoreworkers, shows how they organized within the framework of the fishermen's union (the UFAWU), and explains how as a consequence their numbers steadily shrank until today they represent only a small portion of the labour force. She looks at factors contributing to the destruction of First Nations culture and economy, such as the displacement of aboriginal peoples from key fishing sites and from working in the salmon canneries, and examines the structure and patterns of Chinese and Japanese immigration and the development of the capitalist class and the white working class.

Cheap Wage Labour situates the history of B.C. shoreworkers within the much larger and complex historical enterprise of industrialization, patriarchy, colonialism, and imperialism and provides keen insights into the current fisheries crisis on the West Coast.

ALICJA MUSZYNSKI is associate professor of sociology, University of Waterloo.

Cheap Wage Labour

Race and Gender in the Fisheries of British Columbia

ALICJA MUSZYNSKI

McGill-Queen's University Press
Montreal & Kingston • London • Buffalo

ISBN 0-7735-1376-0

Legal deposit second quarter 1996
Bibliothèque nationale du Québec

Printed in Canada on acid-free paper

This book has been published with the help of a grant from the Humanities and Social Science Federation of Canada, using funds provided by the Social Sciences and Humanities Research Council of Canada.

McGill-Queen's University Press is grateful to the Canada Council for support of its publishing program.

Canadian Cataloguing in Publication Data

Muszynski, Alicja
Cheap wage labour: race and gender in the fisheries of British Columbia.
Includes bibliographical references and index.
ISBN 0-7735-1376-0
1. Labour theory of value – British Columbia – Case studies. 2. Fishery processing industries – British Columbia – History. 3. Fisheries – British Columbia – History. I. Title.
HD8039.F6752C3 1996 335.4′12 C96-900103-7

Typeset in Sabon 10/12
by Caractéra inc., Quebec City

In memory of my father,
Stanislaw Muszynski,
who always believed in me

Contents

Acknowledgments

While I must take full and sole responsibility for any shortcomings contained in the book, I owe an enormous debt of thanks to the many people who both encouraged me in my work and offered invaluable advice and information. First and foremost, Patricia Marchak at the University of British Columbia acted in so very many capacities in relation to this work, above all as mentor and role model. She was my thesis advisor for the dissertation which served as a starting point to the present study. Pat was also the director of the SSHRCC-funded Fish and Ships project which not only offered me financial remuneration but also provided me with a whole group of colleagues and friends. For three years fish and fishing were our major topics of conversation and argument! Special thanks go to Neil Guppy and John McMullan (both of whom also served on my thesis committee), to Brian Hayward who became a special friend in the process, and to Lyn Pinkerton, Keith Warriner and Steve Garrod. The project's secretary, Jean Marchand, provided us not only with expert help but also kept the project functioning on a daily basis. Wallace Clement acted as a consultant to the project and provided continuous encouragement and support as the work unfolded.

I also owe a special debt of gratitude to Keith Ralston, who initiated me into the art of historiography. He generously shared with me his vast and detailed knowledge of the fishing industry of British Columbia. Alex and Margaret Gordon invited me into their home and discussed Alex's years as the first shoreworker organizer for the United Fishermen and Allied Workers Union. Finally, I also wish to thank all the many people related in so very many different ways to the fishing industry in Steveston, Vancouver's Home Plant, Prince Rupert, and the west coast of Vancouver Island who interrupted their own busy schedules and made time to talk to us about their industry.

When I moved to the University of Regina other colleagues continued to offer encouragement and assistance, especially Bob Stirling, Barrie Anderson, and Salina Shrofel. Phillip Hansen not only introduced me to the work of Hannah Arendt but showed me a particularly creative way to read it. Murray Knuttila, head of the Department of Sociology and Social Studies, demonstrated his support by providing me with release time from teaching so that I could finish the manuscript. At the University of Waterloo, special thanks go to Ron Lambert, head of the Department of Sociology, and Elaine Clark Rapley. One of the tangible benefits of this project has been the formation of a coterie of friends and colleagues whose support has been not only an encouragement but also a demonstration of the solidarity of communal organizations of various kinds, whether in the university or in the "real" world. This book would not have been possible without these various communities, and they demonstrate particularly well the interplay between an individual work and the collective consciousness and solidarity that enabled its creation.

Another community into which I was rather sadly introduced was that of the academic publishing world, where support for this project was found sadly lacking, it would seem due to questions of proprietary rights over who can and cannot write on the fisheries of British Columbia. Pat Marchak once again came forward with practical advice and encouragement and entered the fray on my behalf. I wish to thank Peter Milroy and Jean Wilson at UBC Press for their support and encouragement. Last, but certainly far from least, I wish to extend sincere thanks to Philip Cercone and the Publications Review Committee at McGill-Queen's University Press for deciding to publish this work, and to Frances Rooney for her meticulous attention to editing the manuscript.

Map 1 The First Nations of British Columbia. Distribution of ethnic groups, 1850. (Based on Farley, *Atlas of British Columbia*, 7)

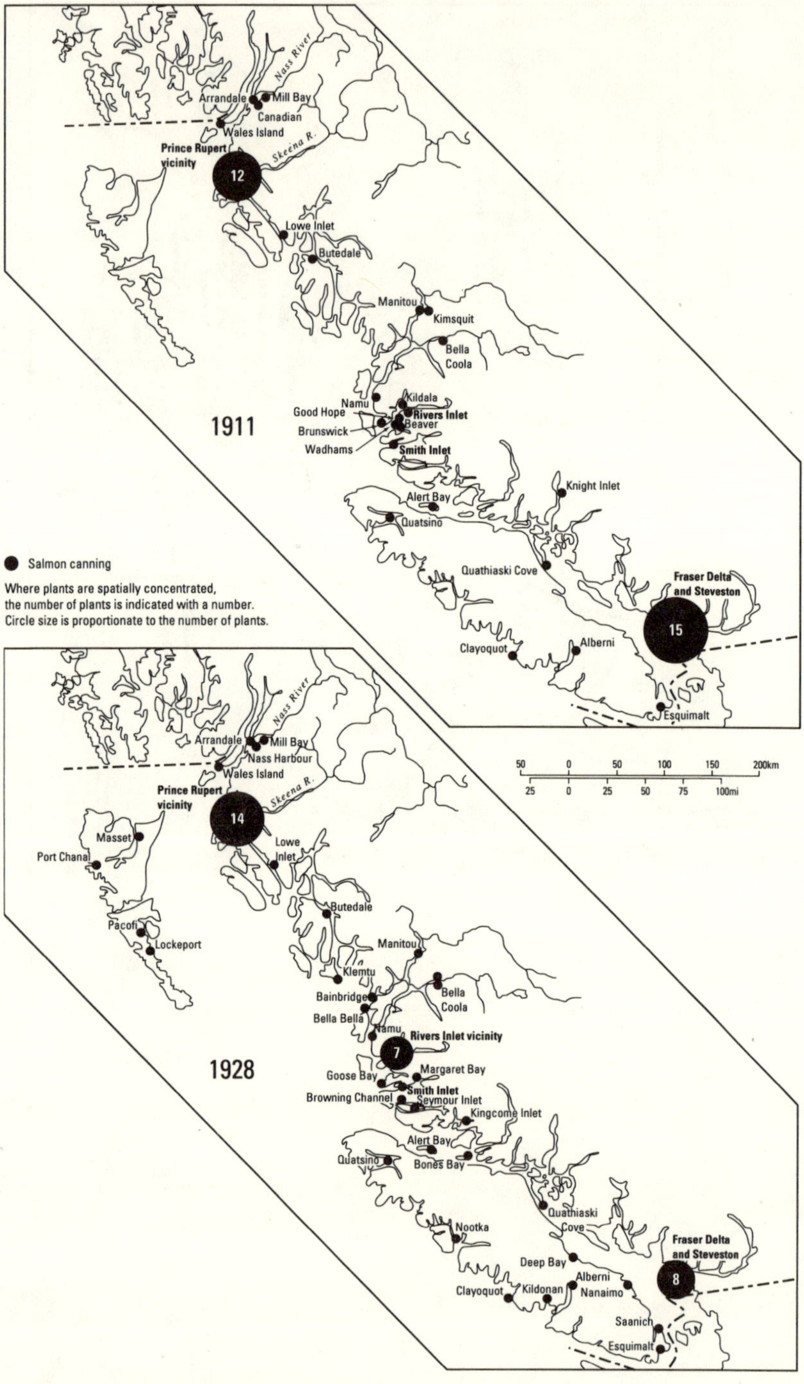

Map 2 Distribution of salmon canneries, 1911, 1928.
(Based on Farley, *Atlas of British Columbia*, 89)

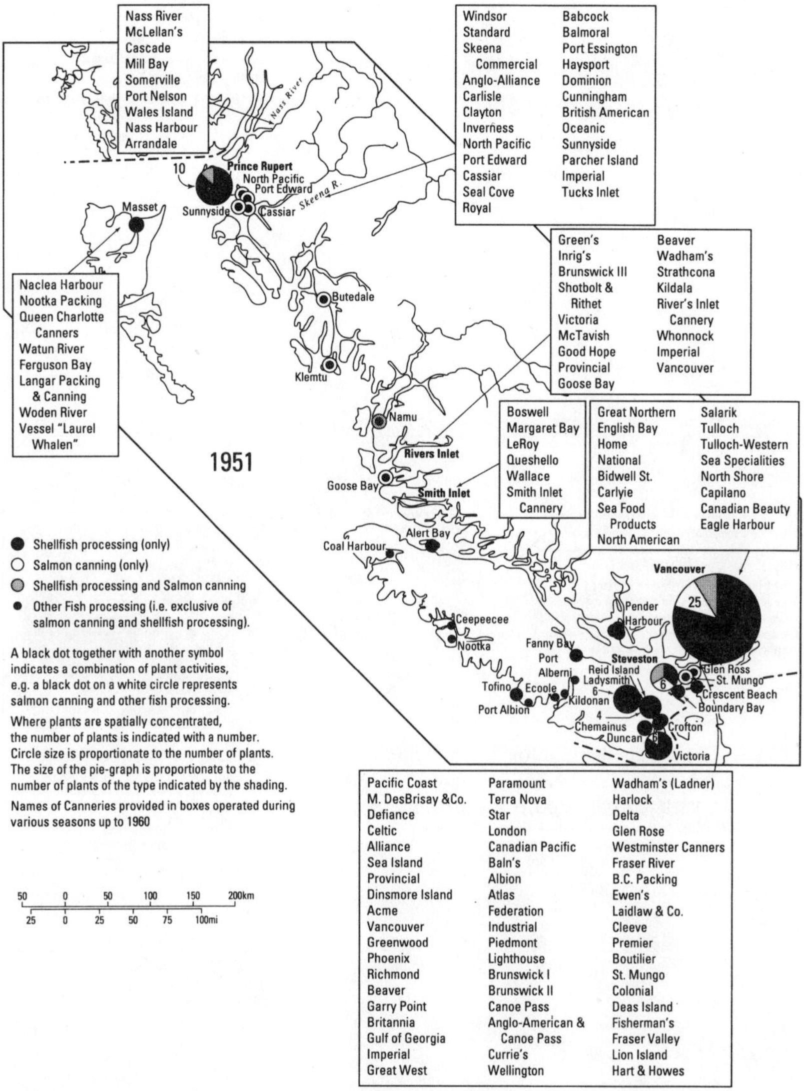

Map 3 Changes in distribution and function of salmon canneries by 1951. (Based on Farley, *Atlas of British Columbia*, 89 and McKervill, *The Salmon People*)

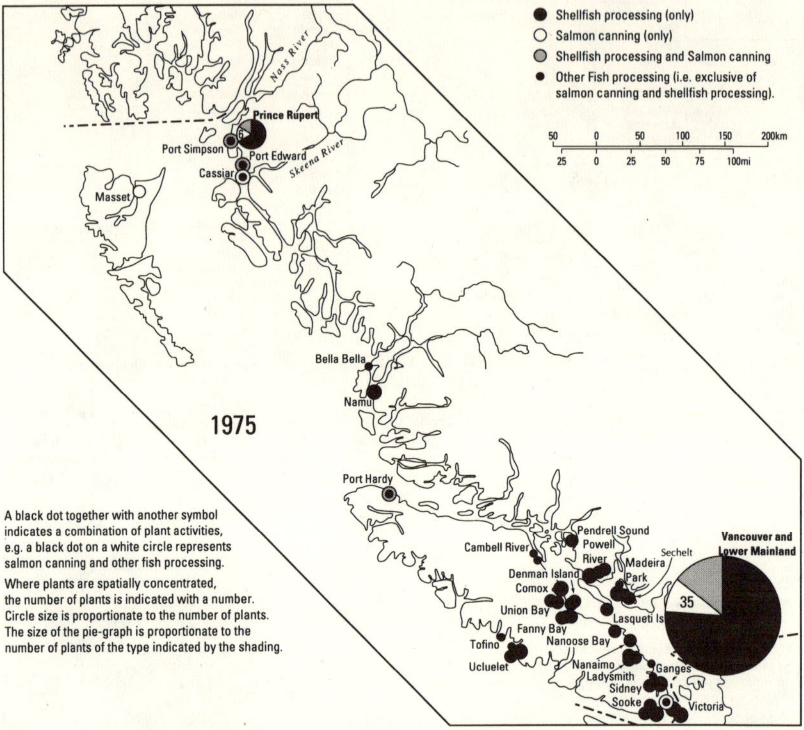

Map 4 Changes in distribution and function of salmon canneries by 1975 as canneries disappeared or were converted into multi-processing operations.
(Based on Farley, *Atlas of British Columbia*, 89)

Cheap Wage Labour

Introduction: The Problematic

Fishing has a certain cachet. As a recreational sport, it brings to mind an image of men "roughing it." Out on the lake in a small boat with lots of fishing gear and bait, they pit their skill and ingenuity against a creature not known for its intelligence but possessing the advantage of being at home in an environment that is foreign to humans – water.

The element of danger and a sense of adventure escalate when the image of fishing moves from leisure activity to a means of earning one's livelihood. Literature is replete with stories of men on the high seas risking their lives in bad weather and tiny craft to capture enough fish to enable them to secure a spartan existence in a coastal village.

Ernest Hemingway hooked his first marlin in 1932, and twenty years later he wrote one of the classics of U.S. literature, *The Old Man and the Sea*. The story takes place in the waters off Havana and reflects Hemingway's knowledge and admiration for "the Cuban fishermen to whom the marlin was a way of life as well as a livelihood" (Waldhorn 1972, 189). But the hero, Santiago, is not simply out on a fishing spree. His journey can also be seen as a heroic quest. "A hero ventures forth from the world of common day into a region of supernatural wonder: fabulous forces are there encountered and a decisive victory is won: the hero comes back from this mysterious adventure with the power to bestow boons on his fellow man" (Campbell as quoted in Garcia 1984, 47). What is particularly powerful in Hemingway's story is that the victory is, indeed, not decisive. But the setting and context are clearly masculine. Gladstein (1986, 47) notes that for both Hemingway and Faulkner "the outdoors served as a masculine refuge, a place of bonding with other men, away from the strong female presence of hearth and home."

The heroic quest is also the symbolic basis for what has been called the "Great American Novel," Herman Melville's *Moby Dick*. Here we enter the no less masculine world of whaling, "a story containing the very prototype of the 'man's world' of fictional setting" (Garcia 1984, 59). Garcia further notes that "the unattainable prize of Ahab's quest, Moby Dick, the leviathan that embodies the enormity of non-human, inhuman creation – the great white whale is, like God the Father, a solitary male" (90).

The themes of Christianity and civilization are also pursued in the work of Melville. Indeed, literary critics have debated Melville's position on racial equality. Karcher (1980, xi) argues that slavery and race are "crucial themes" in *Moby Dick*. The period of Melville's writing, 1846–57, coincided with the "slavery controversy" in the years preceding the Civil War in the United States. Grejda (1974, 10) argues that Melville, in his fiction if not in his own activities, displays a conviction "that the oppressed, regardless of the nature of that oppression, are right to rebel. His militant position carries with it a searing indictment of the white man who has brutalized the nonwhite into an animal or tyrannized him into a fawning subservient – in short, denied the nonwhite his essential humanity."

What these two great American novels illustrate is the often unconscious connection made between fishing and masculine pursuits. They also testify to the symbolism of fishing as an "adventure" in the Western mind. And if the activity of fishing conjures up myriad symbolic and mythic constructions in Western literature, the activities connected to fish once it is brought to shore generally evoke a quite different reaction. Cleaning fish and preparing it for consumption is not celebrated in the literature. Quite often it is assumed to be women's work and associated with the tasks of "hearth and home" that men try to escape, at least temporarily. Thus the symbolic associations around the various activities connected to fishing also reflect the gender dichotomization of various sorts of labour. While men's labour in catching fish has attained heroic proportions in literary classics, the activities associated with fish after it is landed, once it is on shore, far from being celebrated, are often seen as relatively unimportant and, indeed, distasteful. Those who clean fish are not celebrated in literature. Cleaning and processing fish is not a heroic activity.

Using literature to uncover which aspects of a culture are valued involves what Laslett (1976, 325) calls a "theory-laden activity." While my study leaves aside a literary pursuit of the symbolism surrounding the act of fishing, it is a "theory-laden activity" using the discipline of sociology. In particular, it tries to apply an understanding of how Western patriarchy, the colonial and imperial eras, and the extension of the capitalist mode of production to the fisheries of British Columbia shaped the value of the labour used by canners inside the shore plants. What this study tries to unearth are the assumptions canners and those in positions of power who aided them in their industrial pursuits made that allowed them to look at people not as a homogeneous pool of labour but as separate categories each subject to a particular combination of forces that affected its rates of pay. In other words, I treat the concept of labour and the value assigned to it by employers as a "problematic" as defined by Abrams (1982, xv): "A problematic is a rudimentary organisation of a field of phenomena which yields problems for investigation. The organisation occurs on the basis of some more or less explicitly theoretical presuppositions – it is an application of assumptions and principles to phenomena in order to constitute a range of inquiry ... in other words one's problematic is the sense of significance and coherence one brings to the world in general in order to make sense of it in particular." At the same time, I am trying to understand the problematic of salmon canners; that is, their world view that allowed them to undertake specific hiring practices without questioning their assumptions that gender and race were fundamental, unvarying distinctions.

The work that follows is based on a project (my doctoral dissertation) that examined the formation of a waged labour sector, shoreworkers, in the fisheries of British Columbia. Salmon canning began in earnest in the 1870s and represented the first capitalist assault on the provincial fisheries: the commodification of *canned* salmon found a large export market in Great Britain and elsewhere, unlike earlier forms like salted or smoked salmon for which there was limited world demand. From the start, salmon canning involved labour in a factory setting. The assembly line was adopted as the various steps in preparing salmon for canning were broken down and assigned to specific groups of workers who focused on one set of tasks: cleaning the fish, butchering it, making the cans, placing the fish in them, cooking the contents, soldering the cans to seal them, testing for leaks, and lacquering the cans. Some tasks, like can-making, were completed in advance of the salmon runs, as the canneries were being made ready for the season which occurred anywhere from late spring to early fall, depending on the size of the runs that year and on the species of salmon canned (initially, the sockeye were preferred and these run on four-year cycles on the Fraser River). Different groups of workers were assigned to these various tasks and were segregated from one another. "China gangs" (all male) were hired and paid by Chinese contractors to make the cans, to butcher the fish (the machine that replaced them in this process and was adopted early in the twentieth century was called the "Iron Chink" because one machine displaced fifty Chinese butchers) and for tasks like lacquering and testing the cans. Aboriginal women were hired most often to clean the fish and to fill the cans. In a process like can filling, in some canneries the task might be done by Chinese men while in others it was done by aboriginal women. I have not come across any evidence that suggests that these two groups ever worked together. They were separated, even if engaged on the same tasks, and paid according to different systems.

Contractors were sent to recruit whole villages, and the men then worked as fishers while the women and children worked on shore. Aboriginal women also mended the nets used in fishing. When both men and women were employed on this task, again the rates of pay and conditions of employment were specified by gender. White men were employed on an individual basis directly by the canners and filled both the general supervisory positions (the "China gangs" often had their own supervisor hired by the contractor) and the jobs that required overhauling machinery in preparation for the season, tending the machinery in use, and looking after the cannery buildings (for example, the position of watchman). These were the most secure positions, with the highest rate of pay and longest term of employment. Many of the white men were permanent employees, unlike the other groups who were hired at the start of each new season and laid off when it ended. While there was some incentive to keep the China gangs employed for the whole season, once aboriginal villages relocated to a cannery site for the summer, the women were only hired as needed. With the entry of Japanese fishers in the 1890s, and once the men started bringing brides to Canada, Japanese women also began to work in the canneries.

In the early twentieth century, escalating racist sentiment resulted in legislation that effectively eliminated further migration from China to Canada. By then canners had begun cutting labour costs by introducing machinery to the assembly lines and by centralizing plant operations from outlying coastal regions to urban centres, notably Vancouver (the Steveston area of Richmond) and Prince Rupert. They found new labour forces among recently arrived immigrant women with little English whose job opportunities were correspondingly limited.

Salmon canners searched for groups to work inside the shore plants who could be paid the lowest possible wages. In the early days this could be done by paying a contractor rather than bringing the workers themselves onto the cannery's payroll. As the salmon canning industry entered its golden years at the turn of the century, fishers became more militant. Canners tried to apply the same principle of cheap wage labour among their fishing fleets, especially in their hiring of aboriginal and Japanese fishers. However, fishing was one area white men entered in large numbers expecting to maintain a standard of living reflecting the cultural values of the "white" population. Independence and freedom to catch fish on their own terms and to deliver to the highest bidder were important aspects of their definition of themselves as commercial fishers. When the costs of equipping the fishing fleet began to rise with technological developments and adaptations, like gasoline engines on boats, canners were quite willing to turn the burden of outfitting the fleet to the fishers themselves. However, canners continued to press for control over the fishing fleet by keeping an important segment of the fleet directly dependent on the cannery owners who provided boats and equipment on loans which had to be paid back at the end of the season and on condition that the debtors delivered all their catches to the lender.

As Meggs (1991) demonstrates in *Salmon: The Decline of the British Columbia Fishery*, the canners maintained control by playing on the racial divisions within the fisheries and in the province more generally. While Japanese fishers came to compose a significant proportion of the fleet (Meggs estimates that by 1921 they made up over one third of the gillnet and troll fleets [126]), they were shackled by a feudal system that tied them to Japanese contractors and by a hostile white population that made it very difficult for individuals to compete on an even footing in the fisheries. In addition, the canners also kept at least part of the commercial aboriginal fishing fleet dependent by a system of loans and debts tied to fishing vessels and fishing gear. Many canneries, especially those in remote areas, operated company stores that further reinforced this system. Meggs notes that at the turn of the century Japanese fishers were completely dependent on canners. "Since they lived year-round in cannery housing and were dependent on their cannery account for food, fishing gear and even a licence, 'they could not protest against whatever price the cannery might announce'" (50). Cannery housing reflected the racial segregation, with a special type of housing for each group set up in a distinctive location that kept them all physically separated one from the other. While the white employees enjoyed private bungalows on many sites, Chinese men and Japanese men were each put up in bunkhouses reminiscent of the overcrowded conditions found in Nazi concentration camps. Aboriginal peoples camped according to their village groupings, either supplying their own accommodation or being put up in company housing. In 1923 the British Columbia Underwriters Association carried out a survey of existing canneries, and their maps clearly indicate the racial segregation of the various groups in terms of accommodation on the cannery sites.

In the 1890s conflicts among fishers and between them and the canning companies began to escalate and to harden, leading to major strikes at the turn of the century, strikes marred by racial hostility (Ralston 1965). Given the proximity to the United States, it was relatively easy for salmon canners to use U.S.-caught fish, even when the entire commercial fishing fleet in British Columbia was united – a rare event in these early years of racial tension. Only much later, by organizing shoreworkers with fishers, could the latter close down the shore plants altogether. While the United Fishermen and Allied Workers Union (UFAWU), formed in 1945, was founded by fishers and Communist Party organizers, shoreworkers became increasingly militant over the course of several decades.

UFAWU organizers originally focused on recruiting only white men since they held key positions connected to the operation of plant machinery. While these men formed a tiny proportion of the shore plant labour forces, if they were unionized they could effectively close the plants. Chinese men and women sometimes went to recruitment drives and begged to join the union, only to be turned away. The Native Brotherhood began to bargain for aboriginal plant workers, while the Chinese labour force was displaced almost entirely with the formal abolition of the contract system (it persisted in some form through the 1980s). Only in the 1950s did the UFAWU begin to address the issues of racism in cannery agreements; the issue of gender discrimination took even longer, although the principle of "equal pay for equal work" was raised early in the union's history. "In 1968, one of the UFAWU general organizers, Mickey Beagle, presented a brief to the hearings of the Royal Commission on the Status of Women outlining the cumulative results of the use by companies of women as a cheap source of labour. An experienced female general fish worker received 9.3 per cent less than an inexperienced male worker and 24.5 per cent less than an experienced male employed in the same category. A fully qualified filleter received $2.34 per hour (and had to pass tests as well as meet production standards to earn this rate), while inexperienced male help received $2.37 an hour" (Muszynski 1987, 284). The 1973 strike was the first time that the UFAWU addressed shoreworkers' concerns as the top priority, rather than negotiation of fish prices, the most common source of contention. And the strike was fought on the issue not of wages but of the need to introduce a single cannery schedule for all workers rather than the traditional blue lists for men and pink lists for women.

What initially drew me to study shoreworkers was their relative invisibility. At the time (the early 1980s) none of the researchers on the Fish and Ships project was investigating shore work, although Lyn

Pinkerton was studying aboriginal coastal communities which included shoreworkers. Geoff Meggs, former editor of the union paper *The Fisherman*, published his work on the fisheries in 1991, and even he refers very rarely to shoreworkers and certainly does not present them as major participants in the power struggles of the last century. I found it to be a paradox that one sector, fishers, received so much more attention than shoreworkers in much of the scholarly writing. The huge expansion of feminist scholarship in the 1970s and 1980s has resulted in more attention being paid now to labour forces like those in the shore plants of British Columbia.

In 1984 I went to the very inland University of Regina, but I could not leave the topic alone. While I no longer had access to the people and the archives, what drove me was less a need for more detail than the need to develop a "problematic" that would allow me to make sense of the details through a "world view." While the activity of fishers enjoys considerable status in folklore, culture, and literature, the same cannot be said for shoreworkers. And yet, shoreworkers were an essential element in the commercial fisheries of British Columbia, both in the establishment of a salmon canning industry and in the success of the UFAWU.

Classical economic theory argues that labour finds its price on the market according to conditions of supply and demand. I was working within a tradition of political economy, and the work of Marx on the labour theory of value seemed to be crucial. But it did not solve the riddle of the value that seemed to attach as much to people as to their labour power. This paradox set me on the course of combing through feminist scholarship on patriarchy, Marxian analyses of colonialism and imperialism, and studies in the area of race and racism. It was only by trying to integrate concepts and theories from these various disciplines and subdisciplines that I was able to develop a theoretical lens that could account for the way in which salmon canners employed and categorized labour.

My problematic, then, in this work is to develop a general theoretical framework that can help us understand how salmon canners engaged their shore plant labour forces according to criteria of race and gender. In the process, these workers also became proletarianized and eventually were able to utilize the industrial trade union as a vehicle to press for improvements to their working conditions, in which the termination of the use of these categories of race and gender was a central issue. However, trade unionism was only one method used in dealing with the fishing companies. Equally important in how the employers valued labour was the ability of labourers to survive outside a strictly capitalist mode of production. Aboriginal peoples in particular have struggled to retain some measure of control over their traditional economic activities, and fishing was central in the economies of the First Nations of the Pacific Northwest coast. The concerns of the First Nations have at times clashed with those of unionized white fishers. Indeed, the first agreement signed on behalf of shoreworkers was undertaken by the Native Brotherhood, an organization that has at times been anti-UFAWU since it is organized around a broad set of issues of concern to the First Nations and includes aboriginal employers who own fish boats and canneries and employ waged labour and who have historically held important positions of power within the Native Brotherhood. Christian religious values have also been important to the NBBC and have been a source of alienation from union organizers who belonged to the Communist Party.

To the extent that nonmonetary means of subsistence are available to a group to sustain it, wages paid for industrial labour can be negotiated below the rates necessary for survival in a purely capitalist economy, a process traced for the First Nations in chapter 3. Colonial relations had already begun to subjugate aboriginal peoples before industrialization occurred in earnest on their territories and thus played a major role in defining who an "Indian" was and what economic rights such an individual possessed or, more important, did not possess. Also important was the imperial era leading to the defeat of China and Japan and releasing labour pools of men forced to seek work in other countries. These themes are pursued in chapter 4.

In the appendix, the reader will find a specific example of the extent to which the categories of race and gender were reflected in labour agreements negotiated in the late 1940s between the UFAWU (at this point the Native Brotherhood generally signed contracts negotiated by the union for its members, it having been agreed that aboriginal workers could belong to either organization) and the fishing companies organized together as the Salmon Canners Operating Committee (SCOC). The SCOC sent questionnaires to its members, and the tables in the appendix are based on the information provided by J.H. Todd & Sons Ltd for its three canneries during the period 1947 to 1949. The appendix also includes a brief case study of J.H. Todd & Sons, an early entrant into salmon canning and an important company until its absorption in the 1950s by the twin giants in the industry at that time, B.C. Packers and Canadian Fishing Company (Canfisco).

The agreements negotiated between the union and the fishing companies were broken into various "supplements"; in other words, negotiations were made on the basis of the workers' race and gender. The Male Cannery Workers' Supplement covered white men and had the most detailed job classifications and rates of pay. One can see that it covered the fewest cannery workers. The Chinese Cannery Workers' Supplement had fewer categories of labour and the rates of pay reflected the low end of the Male Cannery Workers' Agreement. The fact that the rates of pay for the white men were reflected in monthly increments means that they were given a guarantee that they would be paid for a full month's labour, whether or not work was available. This was an extremely important issue in early negotiations for white male shore plant workers. One can see that in the Chinese Cannery Workers' Questionnaire there are hourly and overtime rates as well as monthly rates, which means that the companies were beginning to depart from their policy of hiring Chinese men and guaranteeing them work for the entire season. Union agreements were not necessarily favourable to this labour force, and indeed by pressing for the abolishment of the Chinese contract system, the UFAWU helped to eliminate this labour force almost entirely from the industry, a process aided by the Exclusion Act of 1923 barring further immigration from China to Canada.

The Women Cannery Workers Supplement had no monthly guarantee. There is little information on wages paid since women were hired on both an hourly and a piece rate basis, which gave each cannery considerable leeway in how it paid these workers. This supplement is further broken down into "White" and "Native." The Networkers Supplement includes four categories: White, Native, and then "Man," "Woman." Only men received monthly guaranteed wages while women were paid by the hour and even then received lower wages than those men on hourly rates. The segregated nature of housing conditions is also reflected in the "Questionnaire re Board, Housing & Services": the first category covered is Male (White), followed by Female (White), Chinese, and last, Natives. Notes that accompanied the completed questionnaires indicated considerable hostility to the Chinese. In answer to one question as to how relations could be improved, someone wrote, "Shoot the bastards!" It was evident that as the companies were being forced to negotiate with the UFAWU, they were planning to eliminate the Chinese altogether rather than integrate them into plant agreements. But the evidence also suggests that Chinese labourers who remained were anything but passive, as the stereotype of their contribution to the industry has often suggested. As early as 1904 they had formed the Chinese Cannery Employees Union to deal with contractors who were paid by the canners and then left for China without paying their crews.

Economists often advance the argument that skill is a crucial component in determining wages, and one often comes across the categories of skilled, semi-skilled and unskilled labour. But these categories also did not make much sense to me when I was trying to understand how canners hired their forces. For example, the "China gangs" would most likely have been categorized as unskilled labour when first employed, and the men did not appear to have taken skills with them to the canneries. But they quickly developed expertise in the areas of can-making and butchering, so much so that in the Royal Commissions of 1885 and 1902 canners acknowledged that these workers were indispensable to the industry. Once they were seen as skilled workers they could then attempt to bargain for higher wages and better working conditions. These demands, coupled with increasingly strict immigration criteria that limited the number of Chinese workers available for salmon canning work, led canners to replace their labour with machines. Thus skill was not a fixed component but one that developed alongside the expansion of the industry and required canners to acknowledge its existence before workers could use it to their advantage in pressing for better wages.

The same could be said for other tasks like filleting. In British Columbia, filleting (except halibut fletching) has always been women's work, and the skill of specific groups of women like the Japanese has been generally recognized. Yet this task is subject to strict quotas and supervision, and pay rates are equivalent to those paid to unskilled men, as Mickey Beagle indicated in her report. While skill is obviously important it is not the only, perhaps in the end not even one of the primary, criteria that determines rates of pay. But race and gender seemed to me, from the evidence I had, to be crucial. Why was this, especially when so much of the standard classical literature in economics and sociology paid so little attention to it?

The following chapters address this question by examining the interrelationship between the concept of labour and the person who provides it. Key here is the notion of value. In a capitalist economy, labour is assigned a monetary value when it is hired in the marketplace. Work (to use Engels' distinction, explained in the next chapter) that remains outside the marketplace thus remains unvalued. This has led to a number of debates within the feminist literature, one of the key ones of which revolves around the domestic labour debate (Hamilton and Barrett 1986). But it is also the case that in most non-industrial societies the work of different groups of people is not granted equivalent value. In the general division of labour, some labour acquires more value than other types of labour (incorporating both "work" and "labour"). For example, in patriarchal societies like those forming Western civilization, the labouring activity of women and children does not appear to have the same importance as that of adult men, even when women perform the same tasks as men. How did these distinctions occur and why? Feminist scholarship in history, anthropology, philosophy, political science, and sociology has expanded greatly in the last two decades, and I use some of the contributions in order to develop an interpretation that can be applied to my specific case study.

In the beginning of the commercial fishing industry, labourers were so invisible as to not even appear on payrolls. Employers paid contractors who in turn were responsible for paying the labour forces they had recruited. Even with unionization, race and gender divisions did not become visible until the labourers themselves began to fight to have the inequalities underlying these distinctions brought forward as union and workforce issues. And even then it took decades of struggle, a struggle that is still unfinished. This is the story I hope to tell and elucidate through a theoretical lens or filter so that the story can be situated in a wider, global, context.

The processing side of the fishing industry of British Columbia was built on notions of race and gender that were embedded, often in an unconscious or semi-conscious way, in the mentality of the province's population of British, Scottish, and Irish background. In addition, the establishment of canneries to process salmon was intimately linked to the global expansion of capitalism. The only reason that these factories were built was to produce a commodity (canned salmon) that could be sold in the marketplace for a profit. Ironically, a proletariat was created in order to feed the British working classes. In his biographical sketch of Jacob Hunter Todd, Anderson (1990, 1063) notes that through "the efforts of the Todds as well as others in the industry, canned salmon became a popular food, particularly in Great Britain, where it was known as 'the working man's feast.'" Creating a commodity to be consumed by the "working man" necessitated holding down costs (including wages), especially given the fact that this was a bulky, heavy item that had to be shipped a long distance. Meggs (1991, 20) notes: "Even after transportation halfway around the world, canned salmon could be sold in Britain for half the cost of fresh meat. Workers would pool their pennies to buy a one-pound can for lunch ... By the end of the century, canned salmon had become so basic to the British diet that it formed part of the army ration."

Readers may find the more ancient use of the term "fisher" in place of the more commonly found word "fisherman" a little strange. After all, in the popular conception it is men who fish, with a few exceptions. But when we look at the history of fishing in British Columbia, there is considerable evidence that, when row boats became a common way of catching salmon, aboriginal men and women fished together. In the latter part of the nineteenth century, these row boats generally required two people to operate them and the gillnets. But the canners only paid aboriginal men for this activity. In 1981 and 82 Janet Burns' (1985) field research on the culture of west coast fishers found that approximately 1 percent of boat owners and 5 percent of deckhands were women. Nevertheless, she also found a strongly defined culture of masculinity among west coast fishers.

Aboriginal women were employed inside the plants, in net mending, and in processing the catches. Aboriginal men hired inside the factories were paid higher wages and generally performed different jobs from those given to the women. When we add the other groups who were hired, it is evident that canners used race and gender as ways to assign jobs and pay different wages. And the activity of fishing itself was defined as men's work. Today this conception is changing, albeit slowly, with more women engaging in commercial fishing as deckhands and as skippers of their own boats.

When we look at wage rates, class relations become intertwined with those associated with race and gender. In creating assembly line tasks, canners wanted to hire labour as cheaply as possible in order to maximize their own profits. Another important consideration for canners was having a sufficient number of employees on hand to do the work. Without refrigeration, fish spoils extremely quickly, especially in hot sunny weather. And fish runs are not evenly spread out over the season, with salmon runs (salmon formed the basis for the industrialization of fishing) varying from year to year. Therefore, labour requirements also vary from one year to the next and within any one season.

Aboriginal women and their children were both the cheapest and most mobile labour force available. The very fact of low wages and short seasons meant that other activities had to be pursued in order to ensure the survival of the village to which individuals belonged. One way canners tried to overcome this difficulty was to send out aboriginal contractors to recruit entire villages to move to the cannery sites for the season. Chapter 3 explores the place of aboriginal peoples in shore work.

There was a core of tasks that canners considered to be indispensable. By combining those tasks, they could guarantee work and pay for the entire season to a relatively small group of labourers. But it was essential that this group remain on site. Chinese men were recruited by labour contractors, who fed the men, paying them at the end of the season, thus binding them to the cannery and their work for as long as canners required their services. The web of relations between these men and the contractors stretched back to China. The peasant economy in Guangzhou (Canton) in southern China was severely affected by Western imperialist expansion in the nineteenth century. Extreme poverty and famine resulted, forcing young men to seek work elsewhere in an effort to stay alive and help their families. The industrialization of the Pacific Northwest, and its small population, created job opportunities for these migrants in many areas, beginning with the gold rush of the 1850s (Barman 1991, 82). Contracting companies offered passage money, charging high interest rates, to North America and helped Chinese male peasants to find work. Canners utilized this system of labour contracting to fill what they came to identify as the most skilled manual jobs: butchering the fish and making the tin cans to contain it. The story of these Chinese labourers, and the related one of how Japanese fishers were also initially hired by labour contractors, appear in chapter 4.

Alongside a discussion of the background of the labour forces that were recruited, it is equally important to discuss the power structures that were developed through the processes of colonization, nation building, and industrialization. Very early in the history of the building of the province of British Columbia, the fisheries were integrated into the global network created by the capitalist mode of production. The potential market emerged as part of the industrialization process originating in Great Britain. The creation of a British proletariat meant that workers were no longer able to feed themselves. A market developed for affordably priced food rich in protein to feed the industrial working classes. Combining this market with the technological innovations associated with canning food, American entrepreneurs developed a British market for canned king salmon. Their Canadian counterparts on the northwest coast followed suit, substituting sockeye for the kings which were not native to the more northern Canadian rivers (Ralston 1981).

The introduction of capitalism was intimately connected with the emergence of British Columbia as a province. While provincial politicians were committed to industrializing activities connected with primary resource extraction (in, for example, forestry, fishing, and mining), they also had a plan for the settlement of the province. Here, the interests of industrialists were not necessarily those of politicians.

Employers sought to hire labour as cheaply as possible. Given the small numbers of European (initially predominantly male) immigrants and the opportunities for self-employment available to these people (for example, as fishers and farmers), those white men who worked for others could command relatively high wages. Men who worked at much cheaper wages were seen to do so because of their racial characteristics. The white population held a popular conception that what distinguished the aboriginal, Chinese, Japanese, and East Indian populations from itself allowed these people to survive on much less, thus undermining and threatening the "white" way of life. Non-white figured neither in the design for provincial settlement nor for the creation of a "white Canada" (Ward 1978). While the aboriginal population had been decimated and was thus not considered to be a threat to this design, Chinese and Japanese immigration at the turn of the century was relatively high, creating racial paranoia in the perceived threat of the "oriental menace" (Roy 1976).

The virulent racism that predominated in British Columbia from the time it was created as a province in 1871 until at least the Second World War, when Japanese Canadians were declared to be "enemy aliens" and interned, is well known. The provincial and federal states played a central role in creating a vision of a "white" British Columbia. As noted, the interests of canners as capitalists sometimes collided with this vision, especially in its more extreme manifestations, as when white B.C. residents and politicians called for the total exclusion of certain races, thus threatening to eliminate the groups that provided canners with cheap labour power. Paradoxically, however, the race consciousness so central to B.C. politics allowed canners to use racial criteria in structuring their labour forces in the first place. The role of the B.C. state in this context forms the theme of chapter 6, and the role of the federal state in creating property relations in the ocean and seas and their fisheries is discussed in chapter 3. The relationships involved in the emergence of the welfare state, unionization, and wage-labour relations in the salmon-canning industry are briefly analyzed in chapters 5 and 6.

While canners, fishers, and politicians are readily seen as active in the important events that shape shoreworkers' lives, the workers themselves are all too often portrayed as victims because of the groups from which they are recruited. These are the groups without power in Canadian society. While much of their history is a story of how they have been rendered powerless, it must not be forgotten that they have actively resisted disenfranchisement and have continuously fought to redress the inequalities forced upon them, not as individuals but as groups denied power because of their race and/or gender.

Much of the struggle of shoreworkers to realize better working conditions has been tied to the unionization of the industry, to the establishment in 1945 of an industrial trade union, the United Fishermen and Allied Workers' Union (UFAWU). The impetus to form an industry-wide union came from fishers and the organizers they hired, many of whom were affiliated with the Communist Party. But trade union status was achieved through shoreworkers, who are paid wages, unlike fishers who negotiate for the price of their catches. In the 1950s legal challenges were launched against the UFAWU, which was charged with being an unlawful combine as defined in the Combines Investigation Act. The challenge centred on the question as to whether fishers were workers who could organize a trade union or independent entrepreneurs (Meggs 1991, 167–74). Important to the UFAWU's case was its claim to bargaining on behalf of shoreworkers who were clearly "workers."

Fishers formed the UFAWU to consolidate their strength during strikes by being able to shut down the entire industry. They could only do this if shoreworkers were organized with them. Otherwise canners could buy fish from non-unionized fishers and U.S. (Alaskan) boats and keep plants open. However, this did not require that *all* shoreworkers be unionized, only those who held key positions in the plants. The UFAWU initially recruited reduction plant workers and machinemen (both groups consisted predominantly of white men). Excluded were Chinese men and all women, who nevertheless insisted that they also be unionized. The gradual incorporation of various groups resulted in the mirroring of racial and gender distinctions in the union's structure. Until recently, for example, there was a pink sheet listing wage scales for women shoreworkers and a blue one for men. And the starting wage for men was often the highest wage for women.

The history of the UFAWU, sketched in chapter 5, has witnessed the attempts of shoreworkers to redress these inequalities within both the union and the industry. Paradoxically, their victories have encouraged canners to further mechanize the industry. As the sexist and racist components in job classifications become eliminated through the UFAWU's initiatives, companies respond by eliminating workers whom they must now pay decent wages. A contemporary example of this process is the introduction of fish cleaning boxes. If adopted on a massive scale, these will eliminate the last manually intensive phase left in the industry, the washing of fish, a job traditionally assigned to aboriginal and immigrant women.

Another severe blow that threatens to eliminate the gains made by the UFAWU is the recent GATT decision that processing fish caught in B.C. waters in the province violates international trade agreements. The UFAWU has prided itself in being able to boast that B.C. shoreworkers are the highest paid fish workers in the world. That achievement has only been made possible because the UFAWU is a province-wide industrial union and has been able to influence federal and provincial legislation to prevent raw fish from being exported outside of Canada to be processed by cheaper labour in places like Alaska or Mexico. For example, in 1989 while American packing plants offered the equivalent of $8 Canadian per hour, B.C. plants paid their workers $14 (McInnes 1989).

While wage labourers and their trade unions struggle to realize decent wages, the larger provincial, national, and international capitalist forces conspire to undermine any gains that are made. It has taken many decades for shoreworkers to begin to redress the racism and sexism that has characterized not only the B.C. fishing industry but also the UFAWU. But in realizing painfully won concessions on the home front, the larger forces have acted to shift the nature and the structure of the industry. One response of companies is to mechanize the work, replacing living labour with the "dead" labour of machines – which are manufactured in places where the labour that makes this machinery is, in its turn, cheaply priced.

The interrelationships of race, class, and gender may be less evident in the provincial fishing industry today. This is due largely to the organized efforts of shoreworkers themselves, together with the alliances they have formed with other groups whose interests they share at least some of the time. But their very success has perhaps spelled their extinction. Canning companies increasingly operate as multinationals, closing marginal plants and consolidating their operations in huge mechanized plants located in Vancouver and Prince Rupert. There was discussion in the fall of 1992 of closing Imperial on the Fraser River, B.C. Packers claiming it can process the bulk of its catches in one centralized operation in Prince Rupert. Since Vancouver has become an attractive location for the monied classes of the Pacific Rim countries, especially people from Hong Kong looking to invest overseas as the colony reverts back to China, B.C. Packers has converted much of its cannery property in Steveston into attractive real estate developments. This is another ironic twist on the interrelationship between class and race. Chinese cannery workers at the end of the last century helped develop the Fraser River fisheries, while at the end of this century the monied classes from Hong Kong are welcomed for quite other reasons than the supply of cheap wage labour. The underlying racism, however, continues with renewed expressions of the need to maintain a "white" British Columbia.

As workers fight the oppression and exploitation embedded in sexism, racism, and class relations, the arena of struggle shifts from the regional to the international stage. Unable to use these distinctions and divisions to maintain cheap labour forces locally, capitalist employers move their operations to other parts of the globe where they can recreate jobs around these divisions. B.C. Packers itself is today a subsidiary of the Weston conglomerate. "The structural weakness in B.C. Packers' and Canadian Fish Company's position in the industry was their vulnerability in forward-market linkages. As early as 1950, George Weston Ltd was engaged in a large, $150 million acquisitions program, part of which involved capturing major food suppliers as a strategy to outcompete other food retail conglomerates. In 1962, Weston secretly purchased a controlling interest in B.C. Packers for an estimated $6 million. By 1974 Weston's had obtained a solid equity interest of 81 per cent for an estimated value of $10 million" (McMullan 1987, 122–3).

The formation of a shore plant labour force in British Columbia can be seen as a case study that fits within this larger global context. It therefore becomes important to try to conceptualize the divisive forces at work and what has allowed them to realize such success worldwide. These are theoretical problems that are engaging increasing numbers of researchers. One of the most promising lines of inquiry to have emerged in this area in the last two decades is that of feminist scholars, whose work crosses disciplinary boundaries and unites such fields as literary theory, philosophy, political science, psychology, history, anthropology, and sociology.

The racist legacy that plagues the history of British Columbia today shocks most people. Canadians like to think of themselves as a tolerant nation, although unfortunately many events in our history contradict this image. What is less evident, although feminists have done much in the past few years to bring it to public attention, is the sexist nature of Canadian society. While people may increasingly be aware that it is somehow wrong to type people and make negative judgments based on the colour of their skin, we still tend to think of gender as a category that fundamentally splits the human species in half. This dichotomization, it is argued in chapter 2, is embedded within Western consciousness. From it, perceptions of difference that engender racism and sexism are made possible. Class exploitation further orders human consciousness in such a way that "difference" becomes conceptually complex, but always categorized in dichotomous ways that carry value judgments of good and bad, of inferiority and superiority.

Chapters 1 and 2 examine some of the recent, and not so recent, work of scholars who have grappled with the theories that relate to class, race, and gender. As the scholarship is vast, it cannot be done justice in the space of a few chapters. However, in order to understand the use of the categories evident in this case study, it is important to ponder some of the more pertinent ideas that have emerged. In particular, it is imperative that we try to bring the theoretical work together so that race, gender, and class can be thought of as interpenetrating relations. That is, one (many traditional Marxists have seen class in this way) of the three cannot be assumed to formulate the other two. The argument in the following chapters is that the three sets of relations as they influenced the formation of the fishing industry in British Columbia originated in historically different periods. Thus it is argued that patriarchal relations, those that order the relations between men and women in dichotomous and unequal fashion, set the stage for Western civilization as we know it today from our history books. Class relations were built into patriarchal relations, changing and further transforming relations between men and women, cementing further layers of inequality. But this was possible partially because the idea that some humans are "better" than others was already part of human consciousness.

While slavery has existed for thousands of years, it acquired a new meaning with Western expansion and the creation of colonies. The historical reality reflected the emergence of a new mode of production, capitalism, out of the mercantilist expansions begun in the fifteenth century. Slavery became commodified within these new sets of class relations. That patriarchal relations did not disappear is evident from the writings justifying racial hierarchies. Many of these used the same arguments previously developed to show why woman was man's inferior and applied them with little modification to non-white populations (Sayers 1982, 96). For example, in an excerpt in chapter 4 from one of the Royal Commissions, canners refer to Chinese male labourers as a "feminine race." Chinese men are described as short, relatively hairless and requiring less food to maintain them than white men. The rationale for characterizing them in this fashion was to substitute their labour power for that of women, who were in scarce supply at the time. They could therefore be paid a scale of wages already established in Europe as suitable for women, who were not judged to need the same amount of food to sustain them as men. This example is but one of the ways in which the three sets of inequalities, those built around race, class, and gender, come together in concrete instances. The theoretical basis for allowing us to see the significance of such concrete examples is the subject of chapters 1 and 2.

Chapter 1 deals very specifically with an analysis of Marx's labour theory of value as elaborated in volume 1 of *Capital* (1967). The introduction to this chapter and its relation to the overall design of the book requires a short digression on the difficulties I encountered in piecing together a history that could be said to relate to shoreworkers in the B.C. fisheries. Those difficulties made me extremely sensitive to the close interrelationship between historical and field research and theoretical formulations that interpret and "make sense" of a vast body of information which in its turn "tests" theories. My difficulty centred around the fact that shoreworkers were often seen by others (for example, federal fisheries officers, investigators conducting Royal Commissions or studies on the fisheries, fishers and the unions they founded, and employers) *and* by themselves as not significant players in the fisheries. Thus there was no one source to which I could turn in order to piece together the events that shaped the lives of these labourers *as* labourers. The conclusion I reached was that I could not make abstractions based solely on the wage labour of these various groups making up the labour force of shoreworkers. A few examples will help to clarify my point.

For example, many of the first labourers were Chinese men, and thus the Royal Commission on Chinese Immigration (Canada 1885) contained information pertaining to their work in the salmon canneries. In fact, canners argued that without this labour force the industry would not have got off the ground. However, the scope of the investigation was much wider, and the role of these men in the fisheries constituted but a small part of the overall investigation. While their significance in the industry was acknowledged, by the end of the nineteenth century they were also seen as no longer crucial to the further growth of the industry. This wider context is extremely important in attempting to understand how and why Chinese male labour was almost totally displaced after the turn of the century.

Compare the example of Chinese male shoreworkers with aboriginal women working in the salmon canneries. While both groups are wage labourers in the same industry, often in the same plants, the contribution of their labour power in the form of wages occupies a totally different place in the framework of their lives and that of their people. While Chinese men were displaced from the industry, aboriginal fishers and shoreworkers continue to this day to play a major role: as workers in the plants owned by large corporations like B.C. Packers, as owners of their own commercial plants in the form of native cooperatives, and as aboriginal peoples with a historic right of access to the food fisheries which form a major part of their subsistence economy. Thus in each of these cases aboriginal shoreworkers belong to another group within which they have occasionally played distinctive roles because they are First Nations peoples and because they are wage labourers. The fact that most aboriginal shoreworkers are women adds a further dimension to the way in which they represent themselves and how they are perceived by various other groups.

During my research it often seemed that nothing in the past of shoreworkers unified them, apart from labour that was for most of them sporadic and temporary. Some of them, especially the new immigrants to Canada, did not want themselves identified with this type of work. On the other hand, aboriginal women I spoke with had an entirely different perception of their work. They took enormous pride in it, but this pride was linked to their long history in which fishing and all the activities connected with it are central to their economy and culture.

All shoreworkers were proletarians. But clearly their labour did not unify them into a cohesive group that could be termed working class. They have been internally divided on at least as many issues as they have been united. While it was in the employers' interests to keep these workers divided against one another, capitalist class interests cannot be taken as the sole explanation, although they are extremely important. There was something about the way in which the canners had constructed the categories of gender and race around specific jobs that mirrored the reality of these people's lives, and that cut across and into working-class relationships.

Clearly, a cannery job for an aboriginal woman who comes from the northern village of Klemtu and is working in a Prince Rupert cannery means something economically different to her from a similar job held by a Japanese woman living in Vancouver and working in a Steveston cannery on the Fraser River. However, through their waged labour these women have developed economic interests different from those of their husbands who fish for the companies in which their wives are employed. And the aboriginal fisher's interests often vary greatly from those of the Japanese fisher. Nevertheless, these conflicting interests have been put aside many times when common interests have served to unite all workers, men and women, regardless of ethnic or racial background, and regardless of the type of work they perform in the fisheries.

While class interests have often united wage labourers, and have also served to unite them with fishers working for the large companies, the basis for a united front has neither always nor necessarily been linked to class relations. The fact that members of the same aboriginal family and/or village or community work together in plants and on the fishing grounds has also been a powerful force, powerful in its economic implications for the family, village, and/or community. Thus men and women have struggled together for the survival of their own people, and have sometimes perceived co-workers as hostile to that struggle. At other times, women, whatever their ethnic or racial background, have formed their own community *within* the plant, and have stood up to the men in their families when these have denied them recognition as important workers and contributors to family income.

Perhaps the only relation that unified all these people since the beginning of commercial fishing is the way in which their labour was being used and transformed through cannery work. All of them were hired to provide labour power that was used to transform raw fish into a commodity (for example, canned salmon) that the companies could sell on the market for a profit. Of the various theoretical discussions on labour, I was initially drawn to Marx's labour theory of value, especially the way in which he tied it into a discussion of the transformation of the British and Western European feudal economies. Here was a theory that encouraged historical investigation and application and that seemed to provide the flexibility needed to incorporate the complexities I was finding in my own research.

While initially promising, however, Marx's labour theory of value could really account for only one very significant part of my project: the capitalist transformation of the fisheries and its implication for waged labour. That is the reason for the focus of chapter 1. It contains a very schematic analysis and critique of Marx's theory. The purpose of the chapter is to introduce the reader to the problematic nature of the very concept of labour itself. It will be argued that Marx himself did not treat labour in a sufficiently dialectical manner. Because he was concentrating on elaborating the laws of the political economy of the capitalist mode of production, he was treating labour almost solely as it was utilized within capitalism. What he missed, and what my research made so glaringly obvious because I was dealing with capitalism in its transitional form, was that labour was already devalued *before* capitalist relations were in place. This is why chapter 2 includes a discussion of the early history of Western civilization and of the establishment of patriarchy as a form of oppression that underlies capitalist class exploitation.

The argument developed in chapters 1 and 2 is that, while capitalism and class relations transformed the fisheries and the relations of the people hired as wage labourers, patriarchy, colonialism, and imperialism must also be studied in order to understand how gender and race were used to structure labour power and to divide labourers against one another. That is, class analysis is not sufficient in studying labour as used in this particular industry. But it does represent a crucial component: the framework for organizing industrial production. For that reason, I begin with Marx's discussion of the transformation of labour power within the capitalist mode of production.

1 Marx's Labour Theory of Value: A Critique

THE CONCEPT OF VALUE

This chapter will explore the extent to which Marx's labour theory of value can be applied to an understanding of how wage labour developed in the British Columbia fishing industry. While a voluminous literature examines Marx's theory, and while I recognize that Marx built his analysis in reaction to the political economy of Adam Smith and Ricardo, my discussion will focus on volume 1 of *Capital: A Critical Analysis of Capitalist Production* (1967), since that work contains a good exposition of how Marx conceptualized the transformation of labour under capitalism.

Central to Marx's argument is the concept of value, the expression of products of human labour in terms of utility (use values), of exchange (exchange value), and/or of both. Himmelweit (1983, 336–7), referring to volume three of *Capital*, quotes Marx's definition of a mode of production as:

> The specific economic form, in which unpaid surplus labour is pumped out of direct producers [and also that this] determines the relationship of rulers and ruled, as it grows directly out of production itself and in turn, reacts upon it as a determining element. Upon this, however, is founded the entire formation of the economic community which grows up out of the production relations themselves, thereby simultaneously its specific political form. It is always the direct relationship of the owners of the conditions of production to the direct producers – a relation always naturally corresponding to a definite stage in

the development of the methods of labour and thereby its social productivity – which reveals the innermost secret, the hidden basis of the entire social structure.

In the capitalist mode of production, relations involving the exchange of products become the dominant form of expressing value; that is, value is attached to the products of human labour. However, both the products and their value are expressions of the capacity of human beings to labour; that is, their ability to add something extra to material reality (value). When value is attached directly to the things (objects) created rather than to the creators (subjects) or to their abilities/capacities, then the possibility exists for the producers (subjects) to become separated from their products (objects). In his discussion of estranged labour in the *Economic and Philosophic Manuscripts of 1844* (1988, 55–67), Marx makes a distinction between estrangement and alienation that is important to keep in mind here. Estrangement is part of the way in which humans are situated in nature and with one another. They produce objects of which they become aware as separate but related to themselves, to nature, and to others. However, with the development of capitalism, estrangement becomes alienation through the transformation of this ability of humans to objectify. Rather than a means of unification, alienation becomes a force that divides humans from their products, from nature, from one another, and from themselves.

For example, consider the processing of fish under the aboriginal mode of production (hunting, gathering, fishing). People expended labour to capture fish and preserve it so that they would have a long-term supply of food. While a division of labour helped develop people's skills in these two areas of capture and processing, at the same time their labouring activity was a bond that knit the group together since the fish was shared communally. Winter festivities acknowledged the importance of nature as provider of the bounty, and the fish was shared with communities that came to visit and participate in the celebrations and the rites. Here labour was estranged, since the fish that was used as food was seen as separate from the original producers/processors, but labour was not alienated. However, when salmon canners hired these same people to undertake essentially the same activities, their labour was now alienated since the employees no longer controlled their labouring activity, no longer related to one another but rather to the employer or his representative, no longer related to nature since fish was an object to be commodified and shipped out of the area, and workers were forced to adopt an instrumental attitude to the objects of their labour.

"Every product of labour is, in all states of society, a use-value; but it is only at a definite historical epoch in a society's development that such a product becomes a commodity, viz., at the epoch when the labour spent on the production of a useful article becomes expressed as one of the objective qualities of that article, i.e., as its value" (Marx 1967, 61). It then follows that labour does not simply involve an acting upon or transformation of material reality (nature). The labourer develops consciousness of her- or himself through the process of labour (as both subject *and* object, since the labourer acts upon material reality but is, at the same time, constrained by it). And most important, the labour process is not a solitary activity. It is cooperative because individuals cannot survive relying solely on their own labour.

Cooperation, in turn, involves division of labour, a crucial concept to be developed later. As long as producers remain in control of both their ability to labour and the products of that labour, social conditions favour cooperation and egalitarian relations. However, when value is socially recognized as residing in products and the products of human labour become detached from those who produce them, then the conditions not of cooperation but of exploitation exist. Products are simultaneously alienated from producers and appropriated by non-producers, precisely because value now adheres to the products (objects) and not to the labour process itself.

Capitalist exploitation involves the further stage of using the objects of labour to exploit the very people who create them, precisely through the predominance of exchange over use values. Specifically, capitalist exploitation involves the separation of producers from the means necessary to ensure their survival (phrased another way, this involves the production and reproduction of labour power). Capitalists acquire control over those means through the institution of private property (private ownership of the means of subsistence encoded in law and enforced through the state). Capitalists acquire control over the products of labour by forcing producers to acquire their means of subsistence in exchange relations, by buying back the products of labour as commodities. Value now visibly adheres to objects in exchange (expressed in monetary form) allowing capitalists to control both the production and the exchange of products/commodities. Returning to the example of the aboriginal economies, however, we see that the process is not necessarily so cut and dried. It will be demonstrated in chapter 3 that capitalism only *partially* transformed the mode of production of the First Nations on the Pacific Northwest coast and that this partial transformation facilitated the development of conditions for cheap wage labour.

"For Marx the value of a commodity expresses the particular historical form that the social character of labour has under capitalism, as the expenditure of social labour power. Value is not a technical relation but a social relation between people which assumes a particular material form under capitalism, and hence appears as a property of that form" (Mohun 1983, 507). As Mohun notes, the value of a commodity expresses the *social* character of labour. Labour is primarily a cooperative activity, otherwise Marx could never posit labour power as socially necessary. As a social process, therefore, it is possible that the products of certain groups of producers can be valued not only in terms of use and exchange criteria, but also in terms of *who* produces them. Although the labour of all members of a society, or community, or household, is socially necessary (in the abstract sense of labour necessary to produce and reproduce society without taking individual labour into consideration), it does not follow that the same *value* attaches to the products if inegalitarian relations exist and are structured by inflexible criteria – such as gender, age, and/or race.

That is to say, there is another manner by which humans have appropriated labour that does not involve exchange value but remains at the level of use values. Division of labour can be both cooperative and exploitive, as Marx demonstrates in his discussion of capitalist class relations. But another form of exploitation precedes and underlies these class relations. I use the term oppression to distinguish this form of inegalitarian social relations from class inequalities, although the two are interrelated and become intertwined in actual relations of production.

In early hominid and human societies, the division of labour between men and women represented cooperative division of tasks for the well-being of the group (K. Gough 1975; Leacock 1981; Leibowitz 1986). But in the historical emergence of Western patriarchy, these cooperative relations were reversed, a process that feminist researchers argue took millennia of struggle and resistance to establish (French 1985; Lerner 1986). The final result was patriarchy, the historical oppression of women by a small group of men who established rule on the basis of qualifications possessed by "man," giving him the right to govern other men, all women and children, and to dominate nature.

As a consequence, "woman" was established as a category that denied humanity to at least half the population. Women were not seen to be human, and became identified with a concept of nature that established material reality as itself something to be dominated by "man." Woman became part of the sphere of material reality to be dominated, entailing her oppression, not by *all* men, but by the system of rule that was established by some men and coincided with the emergence of the state in Western civilization (the origins of which have been traced to the Athenian *polis* and to the Roman code of law).

While this discussion of patriarchy and the oppression of women will be taken up in far greater detail in the next chapter, it needs to be introduced here to point to an important omission in Marx's labour theory of value. For if the majority of women become tied to a category that in its turn deprives them of their humanity and renders them animal-like, the material implications are that the labour women do (tasks clearly differentiated as the work of women through the historical development of the division of labour prior to the emergence of patriarchy) is not human. Although it clearly has use value for the group (generally the household), it receives no *social* value.

Thus, before the further devaluation of labour power that occurs within the capitalist mode of production, labour itself has already been devalued, according to whose labour is used. Capitalist employers were quick to seize upon this devalued (or unvalued) labour from the beginning, as both Marx (1967; 1973) and Engels (1977; 1981) documented. However, Marx explained the extreme exploitation of women and children in the factories as due to capitalist class relations. Although capitalist class relations *further develop* relations of patriarchal oppression, that oppression cannot be understood *solely* by means of class analysis.

In Western Europe before the Industrial Revolution began to transform class relations in the latter half of the eighteenth century, not only did men and women engage in different types of labour, producing different but socially necessary products, but "women's work," that is, the labour process of the female gender, was not valued in the same manner as the labour process of men. In particular, women's labour and the products of their labour were connected to the most important distinguishing characteristic between males and females, their different functions in biological reproduction. The specific tasks associated with the biological production and reproduction of labour power were designated "women's work" and thus without value because they were tasks associated with biological rather than social or cultural functions. And the biological products of women's labour in the birthing process were also without value unless and until they were appropriated by men and inserted into the social, political, economic, and cultural spheres; that is, into society. Thus the relation of both women and children *in* society is mediated by men, a historical process that precedes the emergence of capitalism but that also already possesses a logic of exploitation within which capitalist relations can develop (O'Brien 1983).

Patriarchal oppression is further mediated by the slavery that emerged with colonization and imperialism, during the mercantilist phase preceding industrial capitalism and enabling it through the enormous amassing of wealth that could then be transformed into capital. Although slavery existed prior to this era, it came to represent something unique within the unfolding of the capitalist mode of production: the commodification of human beings rather than of their labour power. But Marx once again conflated the two, arguing they were part of the same process.

In fact, because more labour can be extracted from human beings when they are made to *sell* their ability to labour for a wage (that is, when they are forced to become the agents of the commodification of their labour), Marx argues that the proletarianization of labour is the more exploitive relation. In the end, he blurs the distinction between proletarianized labour and slave labour, as indicated in the passage below. And patriarchal oppression is also seen as part of the same general process. "Whilst the cotton industry introduced child-slavery in England, it gave in the United States a stimulus to the transformation of the earlier, more or less patriarchal slavery, into a system of commercial exploitation. In fact, the veiled slavery of the wage-workers in Europe needed, for its pedestal, slavery pure and simple in the new world" (Marx 1967, 759–60).

It is clear in this quote that Marx recognized the especially brutal character of child labour in England and of slavery in the United States. But he saw slavery as tending toward the proletarianization of labour power itself, regardless of who was in possession of it (whether a British child, an African woman, or a European man). All prior distinctions connected to labour tend toward its commodification as labour power, to which Marx connects his concept of surplus value. Pre-existing relations of oppression, whether in the European family or in the colonies, are seen as leading to this purely capitalist form but are not acknowledged as also determining the new form of exploitation, leading to a different valuation of labour power *within* capitalist class relations of exploitation.

The oversight is partially due to Marx's focus on only one type of wage labourer, European men. Although he cites numerous examples of the exploitation of women and children in the factories, in the end the proletarian condition curiously belongs to men. The following passage clearly illustrates both the force of Marx's dialectical treatment of the capitalist system and his shortsightedness in failing to see that women and children are part of that process but are situated differently in it from men.

Within the capitalist system all methods for raising the social productiveness of labour are brought about at the cost of the individual labourer; all means for the development of production transform themselves into means of domination over, and exploitation of, the producers; they mutilate the labourer into a fragment of a man, degrade him to the level of an appendage of a machine, destroy every remnant of charm in his work and turn it into a hated toil; they estrange from him the intellectual potentialities of the labour-process in the same proportion as science is incorporated in it as an independent power; they distort the conditions under which he works, subject him during the labour-process to a despotism the more hateful for its meanness; they transform his life-time into working-time, *and drag his wife and child beneath the wheels of the Juggernaut of capital.* (Marx 1967, 645; emphasis added)

The active force here is clearly the male working class. Marx does not develop the point that oppressive relations embodied in patriarchal family relations (whereby the male head controls the labour of his wife and children) structure the availability and value of labour power to capitalist employers differentially. Although the commodification of labour power (forcing labourers to treat their ability to labour as an object to be exchanged for means of subsistence) exploits all workers, the "free" labourer (that is, the citizen recognized as such by state legislation) retains responsibility for the production and reproduction of his own labour power. This is not the case for those who are themselves already "owned" through a prior system of oppression (also enshrined in law).

The slave owner assumes responsibility for the production and reproduction of the labour power of his slaves, as does the husband/father for his wife and children. In fact, the European male working class used this privilege to bargain over the price of its own labour power, arguing it was necessary to produce and reproduce more than the costs of its own labour power (the concept of the "family wage"). If men collectively within society are responsible for the production and reproduction of the labour power represented by the members of their families, then the exchange value of their labour power incorporates the family unit. And because men bargain with employers on the basis of their responsibility for their wives and children, the labour power of the latter is valued and exchanged at costs below those necessary for its production and reproduction.

It is interesting in this context, when we turn to the industrialization of the fisheries of British Columbia, that early European settlement of the territory was heavily male, as discussed in chapter 3. Unable to access the devalued labour of white women and children, salmon canners turned to non-white groups, male and female, and applied the same type of criteria, arguing, for example, that Chinese men were a "feminine" race. Examples are also presented in chapter 3 of the arguments made in favour of eliminating the Chinese altogether since it was felt this would promote the immigration of white women and children who could then take the jobs filled by the Chinese male population and further aid in the establishment and development of a "white" British Columbia.

THE COMMODIFICATION OF LABOUR POWER

The difference between the way Marx employs the term value and the dimension added in the present analysis can be traced to the process whereby labour power is itself commodified. "Value is not something intrinsic to a single commodity, considered apart from its exchange for another, but rather reflects a division of labour of independent commodity producers, the social nature of whose labour is only revealed in the act of exchange. Value therefore has a purely social reality, and its form can only appear in the social relation between commodity and commodity" (Mohun 1983, 509).

Marx uses the concept of value to demonstrate that, historically, exchange relations widen the sphere of social relations. When production remains confined to meeting immediate needs, then social organization need not be extensive. When exchange networks begin to be established, as long as production is still primarily oriented to fulfilling the needs of the local community, exchange relations tend to occur only at the social boundaries between communities or are undertaken by marginal groups, like pastoral nomads (Marx 1967, 88; Wolf 1982). However the very process of exchange gradually reorients social relationships until production itself becomes oriented to exchange. This transitional stage (which can occur over many centuries) is complete when products assume the commodity form. Products no longer express utility but are now expressed as equivalents of one another. Money expresses this equivalency, allowing the measure of all other commodities in terms of itself (Marx 1967, part 1).

It is important to keep in mind that this is an abstract argument. Again, the historical record points to a far more complicated scenario before the expansion of capitalism. Indeed Wolf (1982) argues that capitalism itself was able to expand because it was able to seize upon extensive networks of exchange that were already highly developed in the 1400s, before European expansion. Certainly, the First Nations of the Pacific Northwest possessed highly evolved exchange networks despite the existence of at least twenty different language groups in the area that now constitutes British Columbia.

The relation that predominates within a capitalist mode of production is exchange of objects rather than their production. These objects assume value only in the process of exchange or circulation, and that value is a comparative expression (commodities are compared to one another through the prices attached to them). "The value of a commodity can only be expressed after its production, in the use value of another commodity, which, in developed capitalism, is money, the universal equivalent of value" (Mohun 1983, 511).

Production precedes exchange, and more important, the two processes are separated. Value now only becomes visible during the second process, when products circulate as commodities. The money economy, the site of commodity circulation, is separate from the site of production. There is a time lapse between the two processes, and it is through this disjuncture that industrial capitalism develops. That is, before value is attached to commodities, extra value is created and appropriated during production. But value is nothing more than the expenditure of labour on material reality, changing it to meet the needs of subsistence. Exchange relations mask this fact, and this is crucial for the emergence of industrial capitalist production. Since value now attaches to products rather than to labour, it becomes possible to commodify labour itself. The crucial distinction here is between labour (the activity of producers) and labour power (their ability to engage in that activity). As capitalists acquire control over production, they begin to make this distinction, buying not actual labour but simply the ability of people to labour. The result is a complete severance of the labourer from the product of her or his labour, since the employer now dictates how labour is to be expended (Marx 1967, chap. 6). The employer is not concerned with the ability of labourers to survive through production (survival is taken care of on the money market when wages are exchanged for commodities). Rather the employer wants to create something extra for himself, and he can only do this by employing labour, since only labour can create value. Since all the early salmon canners were men, the masculine pronoun will be used in references to employers. While no references to female cannery owners were found, one pioneer canner, John Sullivan Deas, was described as "mulatto," and there were also instances of Chinese merchants directly engaging in salmon canning (Ralston 1972, 1977; Yee 1984, 1986).

Labourers will not voluntarily approach capitalist employers, but must be forced by having their means of production wrested from their control. The only way left for them to survive is to offer their labour power for sale. This sale occurs before the expenditure of labour, and market forces determine its value. Labourers compete with one another to find employment, and their wages become the product of impersonal market forces of supply and demand (the same that determine the pricing of any commodity), but are mediated through class struggle. However, if this were all there was to the process, then there would be no way for the employers to acquire extra value for themselves.

> Our capitalist has two objects in view: in the first place, he wants to produce a use-value that has a value in exchange, that is to say, an article destined to be sold, a commodity; and secondly, he desires to produce a commodity whose value shall be greater than the sum of the values of the commodities used in its production, that is, of the means of production and the labour-power, that he purchased with his good money in the open market. His aim is to produce not only a use-value, but a commodity also; not only use-value, but value; not only value, but at the same time surplus-value. (Marx 1967, 186)

The value of labour power is determined by the costs necessary to produce and reproduce it. That value is not only socially determined (the ability of producers to meet their subsistence needs) but, in the capitalist mode of production, is determined by the exchange relations between wages and commodities necessary for subsistence. Commodities alone are insufficient to assure survival needs, however, because the production and reproduction of labour power takes place outside industrial production. Although this marks another crucial difference between commodities in general and the commodity labour power, Marx fails to analyse the implications of this difference.

> Though labour power appears as a commodity for sale on the market, it is not produced like other commodities. The production of labour power is an aspect of the biological and social reproduction of workers as human beings. This complex process of reproduction involves social relations which are in general different from capitalist or commodity relations. In well developed capitalist societies, for example, labour power is reproduced by household labour which does not receive a wage; in less developed capitalist countries labour power is often reproduced through surviving non-capitalist modes of production. These processes have their own logic and ideology; the pure logic of capitalist relations cannot assure in and of itself the reproduction of labour power. (Foley 1983, 266)

Because a different logic operates in the production and reproduction of labour power, wages alone cannot meet survival needs. Unfortunately, Marx tends to ignore this crucial fact, mainly because he is examining the operation of a pure capitalist mode of production, one in which the production and reproduction of labour power is itself somehow incorporated within capitalist relations.

> The capital given in exchange for labour-power is converted into necessaries, by the consumption of which the muscles, nerves, bones, and brains of existing labourers are reproduced, and new labourers are begotten. Within the limits of what is strictly necessary, the individual consumption of the working-class is, therefore, the reconversion of the means of subsistence given by capital in exchange for labour-power, into fresh labour-power at the disposal of capital for exploitation. It is the production and reproduction of that means of production so indispensable to the capitalist: the labourer himself. (Marx 1967, 572)

COSTS OF PRODUCING AND REPRODUCING LABOUR POWER

At this point Engels' distinction between labour and work is useful: "The English language has the advantage of possessing different words for the two aspects of labour here considered. The labour which creates Use-Value, and counts qualitatively is *Work*, as distinguished from Labour; that which creates Value and counts quantitatively, is *Labour* as distinguished from Work" (Marx 1967, 47ff.). Weinbaum and Bridges (1976, 90) develop the significance of the distinction: "Just as in all societies people work while in capitalist societies people labour, so in all societies people reproduce themselves, but in capitalist societies they consume. In capitalist societies, the market serves as the bridge between the production of things and the reproduction of people." In a pure capitalist mode of production, there would no longer be a distinction between work and labour, because labour would be employed so as to meet all subsistence needs. To date, such a situation does not exist anywhere in the world.

To the extent that work continues, therefore, the costs of producing and reproducing labour power are not only borne in the industrial workplace, through the expenditure of labour, but also take place outside, through the expenditure of work. And while it is possible to attach a price to labour, since it is a commodity, work does not enter the sphere of monetary exchange and thus remains without a price. Marx argues that labour power finds its price on the market through the operation of impersonal market forces, and measures it in terms of the costs of commodities necessary for survival. But there is an added cost that remains unmeasured, and it is precisely this cost without exchange value over which male labourers and capitalists struggle in negotiating wages. Labourers want the full costs of producing and reproducing family labour included in the wage (although there is neither a quantitative measure of that cost nor a standard for how many people are to be included in the family unit, costs involved for each age group, etc.). Capitalists, on the other hand, try to pay wages close to or below market rates, since they use the argument of the "family wage" to price the labour power of women and children below the costs of production and reproduction of labour power. In other words, capitalists turn the arguments used by male labourers against them in pegging wages of a significant proportion of the population below the wages of male proletarians, causing a downward pressure on the men's wages.

Although work occurs outside capitalist production, it becomes dependent on it; for example, the industrialization of housework involves unpaid work in the home, but the objects necessary for work are produced in the factories and must be purchased with wages (Luxton 1980, 1986). In 1986, Statistics Canada calculated the value of unpaid household work as $200 billion or from 32 to 39 percent of the gross domestic product. Approximately 20.7 billion hours were spent on housework, matching almost exactly the 21.2 billion hours spent in paid employment (A. Mitchell 1992, A1).

The crucial point here (and missing in Marx's labour theory of value) is that, to the extent that work continues to produce and reproduce labour power, less labour is necessary, thus allowing the capitalist to pay wages *below* the costs of production and reproduction. As noted, it is here that class struggle also becomes important. For example, in Great Britain, many women and children were legislated out of factories through the successful political organization of male labourers (Lewis 1984, 50, 169–70). Organized working-class men negotiated with employers for a "family wage," thereby cheapening the labour power of women and children. Men viewed the paid labour of women and children as a threat to themselves, and used legislative means to keep both groups out of wage labour altogether, rather than including them in trade union organizations.

Low pay also appears to be a perennial characteristic of women's labour. In modern industrial society, this is often argued to be caused by the assumption that the great majority of female paid labour supplements male income, which in itself is enough to support a family. The notion of a family wage in itself appears to have developed during the nineteenth century partly due to trade union action to improve male pay and conditions, partly due to Victorian 'domestic ideology' which sought to confine women to hearth and home. "A more complex web of power relationships lies behind this, especially given the distance between the idea of the female 'supplementary' and the male 'family' wage, and reality. The reality, as demonstrated for both the seventeenth and twentieth centuries is that many women supported, and support, dependants, while many single men did, and do not" (Charles and Duffin 1985, 19).

But the perceptions of both working-class men and employers were filtered through an ideology that held women and children in low esteem. "Women have traditionally been paid less than men, reflecting both their subordinate position in society and the assumption that women are never the principal source of family income, but are dependent on men as providers" (Safa 1986, 59). For the male working class, consciousness of similarities in the conditions of all people who must labour for wages was obscured by sexist (and racist) ideas concerning the value of the labour of men compared to that of women and children (and also the labour of non-white races). Thus labour power came to be valued using two sets of criteria: one that establishes the value of labour power above the costs of the production and reproduction of the individual labourer, and a second which establishes the value of labour power below those costs (the notion of cheap wage labour).

However, these two sets of criteria emerged historically and do not form part of the logic of capitalism itself (except to the extent that capitalist employers will always seek to minimize their costs of production). Attempts to eradicate the conditions underlying these criteria becomes part of class struggle, within the working class itself and also between classes. During the Industrial Revolution in western Europe and in England, it was European women and their children who were used as cheap wage labour. When European capitalists expanded industrial production to other parts of the globe, they sought to proletarianize the peoples indigenous to the new areas by devaluing their labour power as well. Instead of using a sexist ideology, they were able to employ the racist ideas associated with European colonization. It was more difficult to use gender criteria because women were not perceived as inferior to men in many of these countries, and the debasing of women and children became part of the colonial enterprise (Leacock 1981; R. Reiter 1975). Both Leacock (1954; 1981) and Engels (1981) argue that gender inequality is historically grounded in the transformation of social, cultural, political, and economic relations from cooperation to exploitation. This process can be seen at work in British Columbia with the colonial subjugation of the First Nations and imperial expansion into China and Japan. Each of the three populations was affected in specific ways which had enormous importance to salmon canners who were then able to use all three of these populations as pools of cheap wage labour and, just as important, to have them compete with one another and with "white" labour, leading to the racial hostility among fishing fleets noted in the introduction.

The development of industrial production in European colonies did not necessarily entail the destruction of existing relations of production. In creating cheap labour forces, it was to the capitalists' advantage if the work necessary to human survival could continue apace, as long as it did not interfere with the ability of employers to hire labour power when needed. In other words, relations of production existing prior to industrialization continued but were transformed, made dependent on capitalist exchange relations. For example, in British Columbia, as natural resources were exploited by capitalists or destroyed in the process of industrial expansion, the aboriginal population had a more difficult time acquiring or using those resources for subsistence. Thus it became dependent on commodities that required money or wages. The two means of realizing subsistence could, however, occur together, allowing employers to price labour power below its costs of production and reproduction. Employers could also hire First Nations labourers for short periods of time, an important consideration in resource extraction industries like fishing. The provincial and federal states also acquired important roles in alienating aboriginal control over land and resources and in partially assuming responsibility for maintaining the aboriginal population when it could no longer sustain itself.

SOCIAL AND COLLECTIVE DIVISION OF LABOUR

Social division of labour is present in all human societies. Most, if not all, use gender to assign different sets of tasks (centred around biological reproduction) to men and women. Physiological differences are reinterpreted to enforce social and cultural differences, dividing tasks necessary to human survival such that the tasks themselves become typed as masculine and feminine. As long as little surplus is created, tasks may be gender typed without resulting in one gender's being exploited by the other. Kessler and McKenna (1985) argue that the very ability to "see" gender as dichotomous and mutually exclusive categories is part of Western consciousness. They provide evidence from the anthropological literature to demonstrate other ways of conceptualizing gender that involve more fluid constructions. In certain cultures, for example, biologically defined males and females can become the opposite gender through the tasks assigned to them (if there is a shortage of men, girls can become men by performing masculine activities). The institution of the berdache, found in a variety of different cultures, also indicates the possibility of creating a third gender, a non-gender, or of becoming both genders. Finally, Kessler and McKenna also demonstrate the extent to which biological definitions of sex are also culturally filtered through the perception of the scientific community of biologists as to what constitutes reality.

As Engels saw, the power of men to exploit women systematically springs from the existence of surplus wealth, and more directly from the state, social stratification, and the control of property by men. With the rise of the state, because of their monopoly over weapons and because freedom from child care allows them to enter specialized economic and political roles, some men – especially ruling-class men – acquire power over other men and over women. Almost all men acquire it over women of their own or lower classes, especially within their own kinship groups. These kinds of male power are shadowy among hunters. (K. Gough 1975, 70)

Structured gender inequality stems from the social division of labour, but is historically complete only when surplus is regularly produced and appropriated by a class of non-producers. Between the two stages lies a large time span with diverse transitional forms. In some, like the Iroquois society of the seventeenth and eighteenth centuries, egalitarian relations between men and women prevail as the economy becomes horticultural (Leacock 1981, 150). In others, the status of women is steadily eroded, especially in those societies that practise the "exchange of women," a concept that Levi-Strauss argued marked the establishment of human culture (Rubin 1975, 171–85).

By the nineteenth century the process had clearly been complete for many centuries in western Europe, China, and Japan. These were all patriarchal societies with rigid class systems and in possession of a state apparatus to both enforce surplus extraction and justify class exploitation in ideological terms (mainly through religion). But to say that they were patriarchies is not to argue that they practised oppression of women or class exploitation in the same way. Each developed its own forms of inequality to mirror its own historical practice and in isolation from the rest of the world. Thus, unlike the capitalist mode of production, which is global, patriarchal oppression has been geographically bounded. It is only with capitalism that it too breaks free from regional isolation and becomes meshed with capitalist exploitation.

This process is particularly evident in the colonial expansion of England and France in North America. None of the aboriginal economies had such a stratified system of gender inequality that patriarchal relations could be said to exist. However, the extent of egalitarian relations varied significantly from one group to another. The peoples inhabiting the Pacific Northwest were not patriarchal, nor did they have classes or a state. But the further north one travelled, the more stratified were social relations. The Tlingit, for example, although lineage was traced through the mother's line, had slaves, and Averkieva (1971, 317–42) hypothesizes that these people were developing class relations. Whether or not an economy based on class exploitation would have resulted in the absence of European colonization is impossible to ascertain. Along the northwest coast, there was most likely a mixture of cooperative and oppressive conditions, depending on the history and material conditions of the various groups. The position of women in these social groups thus probably varied considerably, not only from one group to another but also within groups. Anthropologists are attempting to overcome the ethnocentric and androcentric biases that marked much of the extensive research carried out among these people, beginning with the work of Franz Boas.

Whether or not egalitarian relations are a feature of the social division of labour, it is characterized by the producer's controlling the making of her or his products. That is, the producer is responsible for all the operations involved in making particular articles, although a number of producers might share specific parts of the process. This is true even in patriarchal societies, since the head of the household controls the products of those subordinated to him through the ownership of their person, indirectly giving him control also over their labour. But they remain responsible for the products of their labour, to which they cannot claim ownership but whose production is in their control.

Here lies the crucial distinction between the social and the collective division of labour, which Marx had already identified in his early writings. In the collective division of labour, it is not the labourer who is appropriated but the products of his or her labour. In the *Economic and Philosophic Manuscripts of 1844*, Marx distinguishes between the estrangement of objects of production (their objectification by means of which the producer realizes her or his relationship to nature and to other producers) and their alienation. Through alienation in the capitalist mode of production, objects confront the producer as alien things that dominate the producer and render both nature and social relations as forces that control him or her. Even the labouring activity appears to be a force that controls the producer, since the labourer must alienate his or her labour in order to survive.

However, Marx argues in the *Economic and Philosophic Manuscripts of 1844* (1988, 60–2) that this alienation, which he identifies in these early writings with the emergence of private property, is necessary in order that "man" fully realize his species-being. And in so doing all of nature becomes his "inorganic body." "In estranging from man (1) nature, and (2) himself, his own active functions, his life activity, estranged labour estranges the *species* from man. It changes for him the *life of the species* into a means of individual life. First it estranges the life of the species and individual life, and secondly it makes individual life in its abstract form the purpose of the life of the species, likewise in its abstract and estranged form (Marx 1988, 61; emphases in original).

Thus the very process of alienation brings all of humanity and all of nature within the purview of human activity. The fragmentation of humanity through the isolation of the individual is accompanied simultaneously by the universalization of nature and of humanity. The capitalist mode of production is global in its incorporation of humanity and of nature. This process also revolutionizes the division of labour, transforming it from its social to its collective form.

Industrial capitalist production is marked by the separation of producers from their products. The capitalist employer takes control over how products are made. His concern is to employ the means of production (all the inputs necessary to produce commodities, including labour) as cheaply as possible and to increase the margin between input costs and sale (exchange) of commodities. The two types of division of labour are amalgamated, but "the division of labour in production develops at the expense of the social division of labour. At the same time, production in particular labour processes is broken down into its constituent elements, each becoming a separate production process; in this manner the social division of labour develops at the expense of the division of labour in production. But the forces of production developed by capital increase at such a pace that both divisions of labour expand, continually demarcating and revising the lines drawn between them" (Mohun 1983, 132).

It is significant that in England the first products to be reorganized in factories were those associated with "women's work." As Marx illustrates throughout the first volume of *Capital*, women and children formed the bulk of the labour forces employed at the beginning of the industrialization process. As machines came to control the movements of the human hand, tasks were broken down into component parts requiring little skill or physical strength (Braverman 1974, 185–6; E. Reiter 1991, chap. 6). This process also allowed the employer or manager to assume control of the labour process. The craft worker was gradually replaced by the collective labourer. Only the capitalist employer and his executives knew in entirety how the commodity was manufactured. Not only the means of production but also the forces of production (especially the knowledge components) were appropriated from producers.

The distinction between craft and industrial labour was also important to the labour disputes in the provincial fisheries. As long as fishers insisted on craft distinctions, divisions along gear type (gillnet, seine, troll, etc.) made collective organization difficult if not impossible. One of the first groups to promote industrial rather than craft unity in the fisheries were the Industrial Workers of the World (IWW), who also fought racial categorizations. Meggs (1991, 94) quotes an IWW broadside that proclaimed: "Let no nationality or anything else get between you and the price of your fish ... It makes no difference whether you are a Japanese, Chinese, Italian, Indian or Britisher, the bosses rob you all alike. You all belong to one nationality, the working class. The boss is the foreigner." Meggs notes that in the spring of 1913: "The Wobblies, riding a wave of labour discontent then sweeping the province, were offering a new kind of industrial unionism that embraced every part of the workforce, regardless of race or place in the production process" (94). Unfortunately, it would take several more decades before industrial unionism took hold in the provincial fisheries.

As industrial production expanded, the labour involved in the products men had traditionally made was also appropriated and reorganized within factories. Here the concept of skill becomes important. Male craft workers tried to retain control over both their products and their labour power by controlling the knowledge necessary to production. They were better placed than women or children because they had developed organizations or guilds in feudal times to control the entry of workers into specific trades and to restrict the knowledge needed to practise a particular trade to guild members by means of lengthy apprenticeships (Dobb 1978, 116–17). Men came to control specific crafts by employing legal sanctions to control entry and membership. Crafts designated as "women's work" were not protected in this manner, not because they did not involve special knowledge, but because women did not have access to legal protection and restriction of their trades, and because they were excluded from the developing market exchange networks. For example, Alexander (1979, 72–3) provides evidence of the exclusion of women from the old guild crafts in London since the fourteenth and fifteenth centuries. She notes that women's work in the 1851 Census fell into four major categories: domestic and household labour; child care and training; distribution and retail of food; and manufacturing skills based on the sexual division of labour found in the household before the introduction of capitalist industry. These distinctions also applied to the way labour was assigned and paid in the fisheries. Since not enough women could be found to perform all the tasks associated directly with processing fish, a male Chinese labour force was recruited. White men occupied all the jobs involved in tending machinery, a craft occupation involving, in some cases at least, an apprenticeship and certification as in, for example, the various categories of "Machine Man" (appendix). These were the first positions to be unionized by the UFAWU.

For example, in the transition to factory employment, cottage industries increasingly came under the control of middlemen contractors, who supplied cottagers with necessary supplies and collected the finished article. Their profit was realized in the difference between the costs they incurred in supplying cottagers and the sale of the finished articles. Dobb (1978) notes that sometimes cottagers were able to assume this middleman function. These were men who initially disposed of their own labour as well as that of their wives and children. "The important influence in determining the degree to which the domestic producer became dependent was probably the producer's own economic status rather than the proximity or distance of the sources of raw material supplies. And here it is probably true to say that it was the possession of land that was the basis of such independence as the domestic craftsman in this first period of capitalist production retained" (149–50). Middlemen contractors were also crucial in the fisheries and were used to recruit entire groups of labourers, whether "China gangs" or aboriginal villages. And possession of land was crucial in this case because the Chinese labourers were totally dispossessed, unlike the First Nations who were struggling to lay claim to their land and fisheries.

Possession of land was in turn a factor determined by the ability of peasants to acquire control over land and make a success of capitalist agriculture, becoming yeomen farmers. In feudal and commercial agriculture, however, it was men (with rare exceptions) who controlled agricultural production as well as title to the land. E.P. Thompson, in his classic study of the British working class, also demonstrates that the struggle involved men. In fact, women are cited in Thompson's study as important players in the proletarianization of British labour only in contexts like the following: "But this conflict between the artisans and the large employers was only part of a more general exploitive pattern. The dishonourable part of the trade grew, with the displacement of small masters (employing a few journeymen and apprentices) by large 'manufactories' and middlemen (employing domestic outworkers or sub-contracting): with the collapse of all meaningful apprenticeship safeguards ... and the influx of unskilled, women and children: with the extension of hours and of Sunday work: and with the beating down of wages, piece-rates and wholesale prices" (Thompson 1979, 285).

Marx notes two stages in industrial production: machinofacture and modern industry. "In manufacture, the revolution in the mode of production begins with the labour-power, in modern industry it begins with the instruments of labour" (Marx 1967, 371). The proletarianization of women and children took place in the first stage. However, proletarianization also involves men, although acquiring control over their labour power proved more difficult. The proletarianization of male craft workers involved the acquisition of control of the knowledge and tools necessary for the practice of their crafts. This marked the second stage of industrial production, when knowledge was removed from the labouring activity and incorporated in the products of their labour, in machinery.

Along with displacement of knowledge, male craft workers were replaced by women and children when labour involved the simple tending of machines. Craftsmen retained some control, developing trade union organizations to protect their skills from further erosion, in those industries where craft knowledge and skill remained important. Class struggle can be characterized here as a three-pronged struggle: between capitalist and craftsman, between the craftsman and his wife and children, and between capitalist and women and children employed as cheap wage labour. Marx argues convincingly that where abundant cheap labour is available, there is little incentive to mechanize production.

> Before the labour of women and of children under 10 years of age was forbidden in mines, capitalists considered the employment of naked women and girls, often in company with men, so far sanctioned by their moral code, and especially by their ledgers, that it was only after the passing of the Act that they had recourse to machinery. The Yankees had invented a stone-breaking machine. The English did not make use of it, because the "wretch" who does this work gets paid for such a small portion of his labour, that machinery would increase the cost of production to the capitalist. In England women are still occasionally used instead of horses for hauling canal boats, because the labour required to produce horses and machines is an accurately known quantity, while that required to maintain the women of the surplus-population is below all calculation. Hence nowhere do we find a more shameful squandering of human labour-power for the most despicable purposes than in England, the land of machinery. (1967, 393–4)

All of the trends discussed here are evident in the case study that follows. The fishing industry in British Columbia passed through both phases of industrialization, but the process was not an even one. Because tasks were assigned to different groups according to the criteria of race and gender, the availability of a particular group to labour was important in determining the extent to which mechanization was developed. For example, with racist legislation at the end of the nineteenth century, the entry of Chinese male labourers into Canada was increasingly restricted. Those in the salmon canning industry could thereby command higher wages, and this in part led to the invention of the only machine that was developed exclusively for the industry (as opposed to machines that were adapted for use from other industries): the iron butcher (Stacey 1982). Skilled craftsmen (with engineering skills) were hired to maintain and overhaul these machines. The adoption of refrigeration and erection of reduction plants increased the size of this labour force, although numerically it remained small compared to the seasonal labourers hired to process the catches. But these European men were the first to be organized into an industry-wide union that included fishers and shoreworkers.

Thus the UFAWU initially mirrored in its wage contracts the sexist and racist structure prevailing in the industry. Each group had its own agreement, and it was only gradually that all shoreworkers, whatever their race or gender, came to be unionized. This was largely the result of their own demands that they also be eligible for union membership. These developments are treated in detail in the chapters below. The important point for the present discussion is that as industrial production developed in the fishing industry, proletarianization did not affect all labourers in the same manner. Important distinctions were maintained which affected the course of the development of the industry, and of the UFAWU. Class relations were intertwined with the conditions of the various groups of labourers that related to their specific mode of production, and to the patriarchal attitudes of the men who controlled the industry (and, as well, of those who governed the province).

THE CONCEPTS OF SURPLUS LABOUR AND SURPLUS VALUE

Finally, something must be said about Marx's concept of surplus value since this is the key to his labour theory of value. Industrial capitalists employ means of production and the commodity labour power to produce commodities sold on the market. The sale of commodities realizes a value higher than the costs involved in their production. The extra value realized by capitalists (which Marx calls surplus value) can only be created by the expenditure of living labour power – or the transformation of constant capital (including the dead labour embodied in machinery) by variable capital. While constant capital simply transfers value to the commodity, only variable capital can add value to it: "That part of capital, represented by labour-power, does, in the process of production, undergo an alteration of value. It both reproduces the equivalent of its own value, and also produces an excess, a surplus-value, which may itself vary, may be more or less according to circumstances" (Marx 1967, 209).

Surplus value originates in the difference between labour power and its transformation into actual labour activity. To explain how surplus value is created, Marx divides the time labour is employed into necessary and surplus labour time; that is, the labourer works beyond the time necessary to produce and reproduce labour power. In previous modes of production, non-producers appropriated surplus labour directly. Producers were aware which part of their labour belonged to them and which part was taken from them (for example, in the Middle Ages peasants had to work specified periods of time for the lord of the manor and/or had to give up fixed amounts of their produce). Marx calls this the creation of absolute surplus value. In capitalist industrial production another method of extracting surplus value is developed. Labour itself, not simply the length of its employment, is made more productive.

In the first stage, or machinofacture, employers simply extended hours worked without paying proportionately higher wages, thus extracting absolute surplus value from labour. But there are physiological limits to this type of exploitation. As long as there is a large surplus population, there are pressures to keep wages low and reinforcements are available when existing labour forces are exhausted. Ultimately, however, employers using labour in this way will deplete the sources of surplus labour.

A far more effective method, marking the transition to mature capitalism, is to make labour itself more productive by forcing individual units of labour power to create more commodities in a given amount of time. With the aid of machines, less labour time is needed to create the same number of commodities or, conversely, more commodities can be created in the same amount of time. Therefore, the time necessary for labour power to meet its own needs, necessary labour time, shrinks, while surplus labour time expands. At this point, absolute and relative surplus value become indistinguishable, because the working day is not shortened in proportion as labour becomes more productive. "Relative surplus-value is absolute, since it compels the absolute prolongation of the working-day beyond the labour-time necessary to the existence of the labourer himself. Absolute surplus-value is relative, since it makes necessary such a development of the productiveness of labour, as will allow of the necessary labour-time being confined to a portion of the working-day" (Marx 1967, 511).

At the same time, because more commodities can be produced with the expenditure of less labour power, costs of production will fall (once the increased costs of constant capital in the form of new machinery, for example, have been absorbed). If the employer is the sole producer of these commodities, prices may remain at their old levels. More probably, however, competition will force prices to reflect lower costs of production. When the commodities themselves are part of the goods workers must buy to realize subsistence, falling prices will further cheapen the price of labour power, since the same number of consumer goods can now be purchased with lower wages. But if the price of commodities falls, then so does the profit realized by the capitalist. To realize the same profit, the capitalist must produce more goods, which may lead to a crisis of overproduction, which may in turn lower prices further.

When such a crisis occurs, consumption cannot keep up with production. Marxian analysts have studied this phenomenon and developed theories of underconsumption to explain it. Theories based on the falling rate of profit, for example, link the crisis to the problem of falling prices which give rise to the need to create more surplus, which in turn is linked to the need to exploit labour further (for example, by making it more productive than formerly, which will in turn eventually result in another crisis, this time at a higher level of labour productivity). These theories tend to see labour as caught in the contradictions between exploited labour (which is necessary in order to create surplus value and thus profit) and the consumption needs of the labourers themselves. In other words, labourers buy back the products they make, but in order that capitalists might realize profit, labourers must be simultaneously exploited and must be able to consume commodities above production costs. There is a fundamental contradiction at the heart of the capitalist mode of production, and it is this contradiction that crisis theories explore (Shaikh 1978; Wright 1979).

While Marx did see the underconsumption of the masses as a chronic state in capitalist society, it only became a factor in crisis given the dynamics of accumulation and the problem of the rising organic composition of capital. Engels states this position very clearly: "The underconsumption of the masses, the restriction of the consumption of the masses to what is necessary for their maintenance and reproduction, is not a new phenomenon. It has existed as long as there have been exploiting and exploited classes ... The underconsumption of the masses is a necessary condition of all forms of society based on exploitation, consequently also of the capitalist form; but it is the capitalist form of production which first gives rise to crises. The underconsumption of the masses is therefore also a prerequisite condition for crises, and plays in them a role which has long been recognized. But it tells us just as little why crises exist today as why they did not exist before." (Wright 1979, 138–9)

The role of the working class in precipitating crises stems from the development of relative surplus value. As labour is made more productive, less labour power is needed to create the same number of commodities; in other words, dead labour displaces living labour. But dead labour cannot realize surplus value. Less variable capital is needed, which means fewer labourers need be hired. But just as dead labour does not produce surplus value, neither does it consume the commodities it helps to make. Rising productivity can result in both a fall in wages and/or a smaller employed labour force (thus swelling the ranks of the surplus population). Various mechanisms determine which course will actually happen, but the overall effect can be one of underconsumption or overproduction, thus limiting capital accumulation and realization of profit.

Whether or not such analyses provide answers to Engels' question as to why crises occur is beyond the scope of this study. But an important point can nevertheless be made about the role of the working class in such theories, most, if not all of which, posit a pure capitalist mode of production. Even when acknowledgment is made that the production and reproduction of labour power is affected by factors like unpaid work in the home and the existence and exploitation of pre-capitalist relations of production, attention then focuses on the ability of labourers to buy commodities, thus allowing capitalists to realize both their costs of production and a profit. When the demand of labourers for these products falls, then there is the potential for an economic crisis. If capitalism were the only mode of production in existence globally, then the role of the working class in precipitating economic crises through its inability to consume the products of its labour would certainly escalate. One shock absorber for economic crises is surely the ability of labourers to retreat into relations of production outside capitalism, and their ability to ride out these economic storms.

The B.C. fishing industry provides an illustration of the importance of pre-capitalist relations of production as a factor that modifies the potential for crises. Although the market for canned salmon was originally the British proletariat, it was not this segment of the working class that produced the commodity. Obviously, British labourers could not afford to pay high prices for tinned salmon. And clearly, if salmon canners were to pay their labour forces the same rates as those paid to British labourers, there would be no way to make a profit (especially when taking into consideration the high costs of transporting a heavy and bulky article in large amounts to a distant part of the world). It was precisely because canners found sources of labour power other than European male labourers who were immigrating to Canada in the hope of obtaining even higher wages than those earned in their home countries that they could make salmon canning a profitable venture. Chinese and aboriginal canning crews could not afford to buy the commodities they produced, but they were not the market the canners had identified. The market lay elsewhere, and the concern of B.C. salmon canners, as of all capitalists, was to create as large a margin as possible between the sale of the commodity and its costs of production.

Thus to the extent that work proceeds outside the industrial workplace and contributes to survival, labourers can be not only hired more cheaply but they can also weather the effects of economic crises much better than if they were totally reliant on wages and the commodities bought with those wages. Returning to the example of the fishing industry of British Columbia, in the early period canners often allowed aboriginal crews to take those species of salmon inadvertently captured but not usable in canning. A fishery inspector noted in his annual report for 1878: "Of course, from accidental causes, some sacrifice of fish for mercantile purposes has occasionally happened; but, in such cases the cannery proprietors have usually presented the fish gratuitously to the natives around, who have cured the fish by drying for their own consumption." (Canada 1865–1930, Department of Marine and Fisheries [hereafter referred to as DMF], 1878, 292). Such fish had no exchange value for the canners but it had use value as a source of sustenance to their labour forces. While the British proletariat exchanged wages for canned salmon, aboriginal cannery workers continued to rely on their ability to acquire the natural resource outside capitalist relations of production. It should also be noted that by the end of the 1870s crises were occurring because fish landing capacity was outstripping processing capacity during heavy salmon runs. Fisheries officers received reports of massive amounts of rotting fish being dumped back into the rivers because there were insufficient processing facilities and labour forces to handle the runs. By the end of the nineteenth century, fishers could catch far more fish than the canners and labour forces could handle during the peak of the salmon runs in good years.

The discussion here cannot explain how crises in consumption are resolved in the production of consumer commodities by skilled craft labour. Cheaply paid labourers obviously cannot afford such commodities unless precapitalist relations of production enable them to take care of the majority of their needs through unpaid work, thus allowing them to spend their meagre wages on these high-priced commodities (this rarely happens). The movement described here goes in only one direction, production of commodities using cheap labour power for consumption by highly paid labourers, most of them residents in the highly industrialized countries. Many basic consumer goods industries (for example, in textiles and food processing) employ cheap wage labour. An understanding of consumer markets thus involves the recognition that there are different labour forces whose purchasing power differs. That is, there are dynamics at work *within* the working class relevant to an understanding of economic crises.

CONCLUSION

The critique of Marx's labour theory of value developed in this chapter points to both the strengths and limitations of Marx's conceptualization of the centrality of labour in the development of the capitalist mode of production. Regarding the workings of capitalism, Marx's analysis is invaluable. But to the extent that other than capitalist forces remain important *within* the capitalist mode of production, it is necessary to extend Marx's analysis of the role of labour.

The method of extending that analysis proposed in this study is an incorporation of the role of patriarchy in defining labour and in structuring Western consciousness in such a way that thinking itself becomes dichotomized. The next chapter pursues and develops the argument, tracing the historic establishment of Western patriarchy. Although the discussion takes us into historical periods not central to the case study, it is important to treat them very briefly.

The men who established the salmon canning industry in British Columbia came from Western European countries, as did the men who governed the province. Their world view and their vision for the future of the province and of the people who inhabited it were shaped by Western European civilization. That civilization arose on the basis of a certain vision of "man," a vision that in turn shaped the economy and the politics of British Columbia. That vision shaped the place of both the various races and of women in these plans.

Thus, salmon canners consciously created a labour force that was based in part on these preconceptions of the place of various groups of people in the economy and in the society that was being created at the end of the nineteenth century. Most of the people who became labourers were non-European, and thus these preconceptions were not part of their own way of understanding their place in the world. Part of working-class struggle and the development of a proletarian consciousness involved these labourers' own articulation of their place in the newly evolving province, a place those in power denied them. While all of this is linked to material reality, that reality itself was in turn shaped by a vision that was both sexist and racist on the part of those who held economic and political power.

An understanding of how labour was employed in the fishing industry of the province thus requires a discussion of the emergence of patriarchal relations in Western European civilization. The shaping of Western consciousness dichotomized the human population into male and female and came to underlie class relations. However, the two sets of relations, gender oppression and class exploitation, must be analyzed as containing their own dynamic. That is, they cannot be conflated so that patriarchy and capitalism come to be seen as the same. These two sets of relations, in turn, also shaped the racism that has been such a prevalent feature of the history of British Columbia.

However, racism also carries its own dynamic, as discussed in chapter 3. The concept of race also underlies discussions in most of the chapters that examine how each separate group was taken into the fishing industry. In this way racism is tied to the transformation of modes of production of non-European peoples. While canners and government officials operated with a patriarchal mentality that was also racist, the implications of their thinking for the various groups of labourers differed according to the economic, political, cultural, and social history of these various groups. And the women in each of these groups were also affected in a way that differed from how the men in their groups experienced class exploitation. However, while women suffered common oppression *as* women, they were also distinguished according to their race. For example, the experience of First Nations women differed from that of Japanese women, although both were exposed to sexist practices on the part of employers. And although white women also suffered from sexism, on the whole employers treated them far better than other groups of women. By means of the case study, I hope to demonstrate the complexity of the ways in which class, race, and gender are interconnected and how they shape wage labour.

2 Patriarchy and Capitalism

To develop the contradictory interplay between the concepts of capitalism and patriarchy based on the arguments of the previous chapter requires a review of some of the most problematic concepts in the work of Marx, specifically the oppositional tension between his views of nature and of history. Where would or does Marx situate "civilization"? And how do his comments on necessity, consciousness, division of labour, ideology, "the woman question," the family, and the state relate to his understanding of the historical mission of the capitalist mode of production as helping "man" realize his true species-being? While Marx dropped references to species-being in his later work, his earlier views on the situation of humans in nature are useful to understanding his development of the labour theory of value. Unlike structural Marxists like Louis Althusser, this "reading" suggests there are connections to be made between the early and later works (Thompson 1978, 1–210).

Merely reviewing the literature on what has been written on these subjects would be a daunting task. But they are important to broadening the concept of labour, and thus partially reworking Marx's labour theory of value so that it can be used more flexibly to explain the utilization of labour in the shore plants of British Columbia. Nicolaus (Marx 1973, 34–8) notes in his foreword to the *Grundrisse* that, in his critique of Hegel's dialectics, Marx was plagued with the question of where to begin. And Nicolaus argues that it was only on the last page of the seventh notebook that Marx found the "proper beginning" – with the commodity. In bringing forward the idea of the commodity, Marx developed the key antithesis between use and exchange value.

Accepting Nicolaus' postulate, this "key antithesis" can also be used here to examine Marx's treatment of the other concepts noted above. By restricting ourselves to an examination of these ideas as they filter through the labour theory of value, we can ask whether Marx was sufficiently dialectical in his treatment of use value as opposed to exchange value. The concept of dialectics is also extremely difficult to define since it has been used in many different ways by, among others, key philosophers like Hegel and Kant. "There are also some forms of Marxist thought which reject the whole notion of dialectical laws, while retaining a looser sense of dialectic to describe the interaction of contradictory or opposite forces" (Williams 1981, 92). It is in this looser sense that I propose to develop the contradictory forces operating within and among patriarchy, colonialism, imperialism, and capitalism as these were played out in the B.C. fisheries. However, in questioning Marx's use of the term in relation to use value and exchange value, the following passage taken from Roy Bhaskar's seven-page definition of dialectics in *A Dictionary of Marxist Thought* throws some light on the crucial significance of the commodity and of use and exchange value. Bhaskar distinguishes between Marx's critical dialectics, which incorporates a "Kantian moment" and which "may perhaps best be regarded as an empirically open-ended, materially conditioned and historically circumscribed, dialectical phenomenology" from Marx's systematic dialectics. "Marx's systematic dialectics begins in *Capital* I, ch. 1, with the dialectics of the commodity and culminates in *Theories of Surplus Value* with the critical history of political economy. Ultimately, for Marx, all the contradictions of capitalism derive from the structurally fundamental contradictions between the use-value and the value of the commodity, and between the concrete useful and abstract social aspects of the labour it embodies" (Bhaskar 1983, 125).

Marx argues in the *Grundrisse* (1973, 158) that the historical role of exchange value gradually leads to the global development of capitalism. "Patriarchal as well as ancient conditions (feudal, also) thus disintegrate with the development of commerce, of luxury, of *money*, of *exchange value*, while modern society arises and grows in the same measure" (emphasis in original). But he treats original production for use value only as "spontaneous." At this point, "human productive capacity develops only to a slight extent and at isolated points" (158).

Production in general is an abstraction, but a rational abstraction in so far as it really brings out and fixes the common element and thus saves us repetition.

Still, this *general* category, this common element sifted out by comparison, is itself segmented many times over and splits into different determinations. Some determinations belong to all epochs, others only to a few. [Some] determinations will be shared by the most modern epoch and the most ancient. No production will be thinkable without them; however, even though the most developed languages have laws and characteristics in common with the least developed, nevertheless, just those things which determine their development, i.e. the elements which are not general and common, must be separated from the determinations valid for production as such, so that in their unity – which arises already from the identity of the subject, humanity, and of the object, nature – their essential difference is not forgotten. (Marx 1973, 85; emphasis in original)

Two important points can be drawn from this quote. First, within the human condition at any given time, we can trace its most ancient antecedents. That is, Marx alludes to a concept of history here that is not linear, but which contains potentiality (a possible future), realization (that which is at any one "moment"), and all previous conditions as mirrored in the present and pointing to the future. Second, difference must be "picked out" from production since production characterizes all human endeavour. That is, all production everywhere and at all times creates use value. It is exchange value that transforms production and, ultimately, use value itself. The complete transformation wrought by exchange value transforms nature and leads to the full realization of "man's" species-being. At this point, history begins. Thus, all previous relations, because they are stamped by "man's" inability to consciously fashion nature and himself, remain rooted in pre-history.

If capitalism represents the global transformation wrought by the development of exchange value and its ultimate commodification of labour power, what is to represent the transformations wrought by the development of use value? The concept of patriarchy can be elaborated to develop an oppositional relationship between use value and exchange value.

In concentrating on exchange value, Marx ignored the potentialities contained within use value itself. As we will see below, Marx connects production solely for immediate use with necessity and with the domination of nature over human beings, a cycle in which they are caught trying to assure group survival without any great development of the knowledge that would later allow them to "master" nature and, with it, necessity.

Recent feminist scholarship suggests that the period between the maturation of the capitalist mode of production and earliest hominid societies is not only vast in terms of time, but also complex in terms of the possibilities for the development of human societies. While industrial society inverts the relation between exchange and use value, production for use and production for exchange are not simply or only mirrored in this inversion. That is, the primacy of production for localized use illustrates a complex variety of social relationships, from cooperative to oppressive, and relations of exchange develop *within* these formations in very complicated ways. To characterize human development before capitalism as "pre-historical" suggests that humans exercised little choice in their approach to production and to their environment, an instance of patriarchal consciousness. Numerous examples in the various writings of Marx suggest he held such a view. For example, in the *Grundrisse* (1973, 107) he traces the elements of modern bourgeois society in various pre-capitalist modes of production, but notes that "mere hunting and fishing peoples lie outside the point where real development begins."

The next chapter provides evidence that Europeans encountering aboriginal peoples in North America were also operating with such presuppositions and that these views then structured the relations of domination imposed on the First Nations by European colonizers, who were thus morally freed from questioning their actions. Indeed, many viewed themselves as harbingers of "civilization" to peoples without religion or culture. Far from acknowledging their destruction of aboriginal economies as morally wrong, they viewed the forced integration of aboriginal peoples into waged labour as part of the "civilizing" process.

Colonial relations were also grounded in pre-existing patriarchal relations of oppression. To use the concept of patriarchy as a way to mark the changes contained within the development of use value requires not only that we focus on labour itself but also on that most primary of human relationships – sexuality. Marx himself was sensitive to the implications of human sexuality in the relations between men and women and in the division of labour.

THE CONCEPT OF NATURE IN THE WORK OF MARX

In the approach to *woman* as the *spoil* and handmaid of communal lust is expressed the infinite degradation in which man exists for himself, for the secret of this approach has its *unambiguous*, decisive, *plain* and undisguised expression in the relation of *man* to *woman* and in the manner in which the *direct* and *natural* species-relationship is conceived. The direct, natural, and necessary relation of person to person is the *relation of man to woman*. In this *natural* species-relationship man's relation to nature is immediately his relation to man, just as his relation to man is immediately his relation to nature – his own *natural* destination. In this relationship, therefore, is *sensuously manifested*, reduced to an observable *fact*, the extent to which the human essence has become nature to man, or to which nature to him has become the human essence of man. From this relationship one can therefore judge man's whole level of development. (Marx 1988, 80–1; emphasis in original)

Unfortunately, Marx does not develop this line of thought, contained in the *Economic and Philosophic Manuscripts of 1844*, except to immediately connect this relationship to the development of private property – as does Engels in *The Origin of the Family, Private Property and the State*. It is, nevertheless, noteworthy that although Marx writes of the relation between *man* and *woman*, the human essence belongs to man (and this is most likely not simply a problem of translation).

In the end, Marx pays little attention to the development of human sexuality and its relationship to the division of labour between men and women in various modes of production. Schmidt (1973, 170) argues that "what Marx called the natural (*naturwüchsig*) division of labour within a tribe or a family is based on differences of sex and age, i.e. on a 'purely physiological foundation.'" He further argues that, for Marx, the division of labour becomes social with the development of relations of exchange: "In this way, the connection between individuals is restored, but as a socio-historical connection" (170).

There is thus a curious contradiction at the heart of Marx's thinking about the labour theory of value. On the one hand, nature is the stuff of humanity. Consciousness of themselves as humans relates people back to their connections with nature, and through it, with one another. At the same time, however, Marx keeps insisting that the development of a historical consciousness that represents humans truly realizing themselves as humans by conquering the forces of nature and making conscious use of them is related to the full maturation of capitalist relations and their dissolution in socialism. There are a number of instances where this contradiction is evident; for example, in this passage taken from *Capital* (1967, 79): "Those ancient social organisms of production are, as compared with bourgeois society, extremely simple and transparent. But they are founded either on the immature development of man individually, who has not yet severed the umbilical cord that unites him with his fellowmen in a primitive tribal community, or upon direct relations of subjection. They can arise and exist only when the development of the productive power of labour has not risen beyond a low stage, and when, therefore, the social relations within the sphere of material life, between man and man, and between man and Nature, are correspondingly narrow."

> Out of the act of exchange itself, the individual, each one of them, is reflected in himself as its exclusive and dominant (determinant) subject. With that, then, the complete freedom of the individual is posited ... The other is also recognized and acknowledged as one who likewise realizes his self-seeking interest, so that both know that the common interest exists only in the duality, many-sidedness, and autonomous development of the exchanges between self-seeking interests. The general interest is precisely the generality of self-seeking interests. Therefore, when the economic form, exchange, posits the all-sided equality of its subjects, then the content, the individual as well as the objective material which drives towards the exchange, is *freedom*. Equality and freedom are thus not only respected in exchange based on exchange values but, also, the exchange of exchange values is the productive, real basis of all *equality* and *freedom*. (Marx 1973, 244–5; emphasis in original)

But if the original relation between man and woman is marked by the inequality and the unfreedom of women with respect to men, then no amount of the development of relations of exchange (even if they culminate in socialism) will *necessarily* eradicate that oppression. This logic stems from the assumption that the original relationship of inequality and unfreedom is *not* physiologically rooted but is itself socially produced by human beings. Central to this development of oppression is a certain view of nature and of woman's place in it that Marx himself was unable to overcome.

Mills (1987, 51) argues that Marx "shifts from the personal man-woman relationship and the specific logic of the family to a mediated relationship of abstract equality between man and woman in the sphere of productive relations." This introduces yet another problem in Marx's thinking, his shifting between the needs of the individual and the needs of the family, touched upon in the previous chapter. Marx tends to dismiss social ties based on kinship. "The less social power the medium of exchange possesses ... the greater must be the power of the community which binds the individuals together, the patriarchal relation, the community of antiquity, feudalism and the guild system" (Marx 1973, 157). He goes on to argue that "patriarchal" as well as ancient conditions dissolve with the development of commerce, of money, of exchange value, "while modern society arises and grows in the same measure" (158). As humans develop socially and create a truly human history, the older relations based on physiological need disintegrate. Patriarchal relations, including those represented in family formations, also dissolve, leaving the individual a truly social being defined by and through "his" relations of exchange with other free individuals.

But this concept of "individual" is a modern one. As Williams (1981, 134) points out, originally the term meant the opposite of its current usage – "indivisible" or connected. "The modern sense of *individual* is then a result of the development of a certain phase of scientific thought [especially Darwin's work with which Marx was both familiar and greatly impressed, wanting to dedicate the first volume of *Capital* to Darwin] and of a phase of political and economic thought" (136). Williams also notes that this concept of individual is tied to the evolution of another important category – "Man." The historical development of these ideas is also connected to the evolving meaning of civilization, especially the meaning it came to have in the nineteenth century.

While Marx never defines what he means by the term "civilization," there are sufficient comments in his work to suggest that he means the development of exchange value. For example, in the *Grundrisse* (1973, 256–7) he writes of the "*civilizing influence*" (and he places a stress on these terms) of external trade. "The degree to which the movement towards the establishment of exchange value then attacks the whole of production depends partly on the intensity of this external influence, and partly on the degree of development attained by the elements of domestic production – division of labour etc." He gives as an example the development of English agriculture in the sixteenth and seventeenth centuries, when it "thus lost the character of labour for use value" (257). Just as the development of relations of exchange enable truly social organization and thus history, production purely for use value situates humans in pre-history and pre-civilization.

It is important, however, to keep in mind the interconnections Marx emphasizes at the very beginning of the *Grundrisse* among production, distribution, exchange, and consumption. While in reality no one of these can be disentangled from the others, Marx's thesis in these notebooks is that when relations of exchange come to dominate and direct production for local use, then humans enter into a new phase of production, one that is globally rather than locally situated.

Because Marx concentrated on the implications of this global development of production for exchange, he paid insufficient attention to the social and historical character of the transformation of production for use. That transformation has been collectively forgotten. Western human history is seen to begin with "civilization," a forced beginning where dating itself begins with Christianity, the birth of Christ, the beginning of linear time and linear thinking. Everything before was either "savagery" or laid the foundation (for example, in Greek philosophy and Roman law) for civilization, for man-made history. In spite of his dialectical understanding of history as it relates to production, Marx was also guilty of such thinking, especially in his insistence on the benefits of Western scientific thought as enabling domination of nature. Schmidt (1973, 37) notes that for Marx, "the essence of man is not something fixed. Man's essence has still not appeared in its entirety. On the contrary, in history up to the present, which is designated as 'prehistory' by the fact that men are not in control of their own powers in relation to nature, the essence of man has been brutally subsumed under the material conditions for the maintenance of his existence." Both Schmidt (1973) and Mills (1987), in their treatment of Marx's approach to nature, refer back to his critique of Hegel. While Marx critiqued Hegel's notions of "first" and "second" nature, he did not in his own thinking overcome the implicit dichotomization contained in this splitting of nature and of human history.

"FIRST" NATURE/"SECOND" NATURE

> Hegel described the first nature, a world of things existing outside men, as a blind conceptless occurrence. The world of men as it takes shape in the state, law, society, and the economy, is for him "second nature," manifested reason, objective Spirit. Marxist analysis opposes to this the view that Hegel's 'second nature' should rather be described in the terms he applied to the first: namely, as the area of conceptlessness, where blind necessity and blind chance coincide. The "second nature" is still the "first." Mankind has still not stepped beyond natural history. (Schmidt 1973, 42–3)

However, the problem for Marx was not so much that reason is elevated and opposed to nature, as that reason *in* nature has not progressed far enough. Marx opposed raising reason to the status of "Spirit," of a guiding force divorced from production, but he did not seem to object to the application of reason, especially scientific reasoning, to an understanding and control of nature. Marx also appears to hold all that comprises "first nature" in low regard, rather than realizing that the split is a social creation in thinking that reflects back on the organization of relations in production.

The term "civilization," like many others that we tend to take for granted as somehow stating the obvious, has had a rather complex and changing history. Williams (1978, 13) notes that the original sense of the word meant "bringing men within a social organization," and that this sense was incorporated within the term "civil society." However, "civilization" was also extended, thanks to the "new historical rationality of the Enlightenment," to mean an "achieved state" that could be contrasted with "barbarism." It was further extended in this line of thinking to mean "an achieved state of *development*, which implied historical process and progress ... It was the crucial step beyond the relatively static ('timeless') conception of history which had depended on religious or metaphysical assumptions. Men had made their own history, in this special sense: that they (or some of them) had achieved 'civilization.' This process was secular and developmental, and in that sense historical. But at the same time it was a history that had culminated in an achieved state: in practice the metropolitan civilization of England and France" (13–14). Williams (16) notes that the central element in the creation of "Universal Histories" was reason, "an enlightened comprehension of ourselves and the world, which allows us to create higher forms of social and natural order, overcoming ignorance and superstition and the social and political forms to which they have led and which they support. History, in this sense, was the progressive establishment of more rational and therefore more civilized systems."

Williams argues that Marx introduced the "history of labour" into this view, thus retaining the importance of "man's" relation to nature within material history. Williams (1978, 19) credits this as the most "important intellectual advance in all modern social thought. It offered the possibility of overcoming the dichotomy between 'society' and 'nature.'" However, Williams notes that later Marxist thinking persisted in "an earlier kind of rationalism, related to the assumption of progressive unilinear development" (19). Its worst manifestation was in that brand of Marxism that insisted on the base/superstructure dichotomy, which resulted in, among other tendencies, economistic thinking. While we can see a different orientation within Marx's writings, there does nevertheless appear to be a certain form of rationalism.

Mills (1987, 55) argues that Marx's thinking on the overcoming of the inequality between men and women is a form of "economic determinism" that contradicts his "own insight that each specific subject requires its own logic." "Marx focuses on the heterosexual love relation as the paradigmatic form of intersubjectivity. With the development of his critique of political economy, Marx concentrates on the relation between nature and history as a historical dialectic in which labor is central to the formation of consciousness. Within this understanding of the dialectic the paradigmatic form of the intersubjective relation occurs within civil society between workers; it is a relation mediated by the process of objectification (object-creating)" (50).

But it is not only in the relation between men and women that Marx falls into a type of economic determinism. This problem encompasses most of his thinking about the employment of labour prior to the development of capitalist relations of production. "Under slavery and serfdom there is basically no division between labour and its natural preconditions. Both moments merge to form an undifferentiated *natural basis* for the existence of the slave-owner or the feudal lord" (Schmidt 1973, 81; emphasis in original).

This type of conceptualization poses major problems for understanding the industrialization of the B.C. fisheries. Applying such an analysis to British Columbia would involve focusing on the industrialization of the province, the development (by way of colonialism) of the capitalist mode of production. All existing relations of production would then be studied in terms of their incorporation and transformation from use to exchange values, by which to generate surplus and profit for the capitalist entrepreneurs. The main dynamic would be the confrontation between a nascent (or immigrant) bourgeoisie and labour. Lost altogether from view would be the dynamic relations that had already for thousands of years structured the aboriginal economies in relation to the same activity – fishing. In fact, European explorers and fur trading posts depended for their subsistence on the ability of the First Nations to keep them supplied with food, especially fish. The main sockeye-producing stream in southern British Columbia is named after the explorer, Simon Fraser, who "relied on native people for food every step of the way. Although he had acquired some salmon before leaving, he also took caches of fish he found along the route" (Meggs 1991, 13).

Much of the analysis in the following chapters traces the dynamic of how labour was employed to create exchange value. However, treating labour only as it is transformed into labour power fails to distinguish between groups of labourers. Thus, from the point of view of a capitalist employer, it should make no difference whether a labourer is aboriginal or Chinese, male or female. After all, those kinds of differences belong to a pre-capitalist division of labour (part of pre-historic "first nature"). But these were exactly the sorts of differences salmon canners used to establish their labour forces, to pay them, and to define the work tasks that had to be performed in the fish plants.

Also lost are distinctions between the relations of production of the aboriginal population in a hunting/gathering/fishing mode of production and the relations tied to an "asiatic" mode of production which created, under imperialism, a surplus population of Chinese male peasants who sought employment overseas in order to help feed impoverished families in China. Precisely such distinctions enabled various groups of labourers to subsist *outside* a capitalist economy, thereby allowing their labour power *within* capitalist enterprises to be priced differently. In other words, these groups did not come to an amorphous labour market where only their labour power was commodified through the operation of the law of supply and demand. Negotiations for their labour, and its pricing, were conducted on a group basis. It is at this point that the concept of patriarchy can be used to extend the labour theory of value.

In what follows, we should keep in mind the problems posed for European entrepreneurs, reared in a culture where patriarchal thinking is so deeply engrained that it subconsciously motivates action, in undertaking capitalist industries in a sparsely populated area of the world with an aboriginal population that is not "civilized" (using the definition developed above). In other words, a central component of state building in British Columbia involved carving out a public as well as a private realm. Neither existed, since the concepts themselves are historically specific. We will examine the concept of patriarchy to explain the dichotomizations that characterize Western thinking, and to throw some light on the categorization of people in light of the labour they provide. But first a cautionary note is in order.

Before the global spread of capitalism, patriarchy as a system of power had existed for millenia in China, Japan, and India. What distinguishes patriarchy as a power system from capitalism as a power system is that patriarchy is localized (that is, the notion of territory is important in dominating and exploiting groups of people), while capitalism is global. However, almost by definition, capitalist relations of exploitation not only take root in territories marked by concepts of power derived from Western patriarchal thinking, but in spreading to other parts of the globe, they connect and mesh with other existing patriarchal systems of power. Extremely complicated sets of relationships of exploitation, domination, and oppression result that require much further study than has been accorded them in Western thought and writing.

In the B.C. fishing industry, for example, the use of Chinese and Japanese labour must be connected not only to the consciousness of the canners, steeped in sexist and racist attitudes that derived at least in part from their own history of Western patriarchal thinking, but also to the histories of patriarchal domination that characterized China and Japan respectively. The later use of immigrant East Indian women as cheap wage labourers further complicates the analysis. It is impossible to provide detail here because this area of scholarship is just beginning, at least in Western countries, much of it by feminist scholars who are beginning to connect with their counterparts in these other countries. (A good example here is the collaborative work of Maria Mies, professor of sociology in Cologne, Germany, with Vandana Shiva, Director of the Research Foundation for Science, Technology and Natural Resource Policy in Dehradun, India.) To think that we in the West can provide a sufficient analysis reflects our own Western patriarchal and ethnocentric biases. At most I hope that this study can open a dialogue that will encourage researchers from the countries that contributed labour forces to the B.C. fisheries and/or that have been affected by the provincial industry to complement this research.

Of course, the need for the voices of the labourers to be heard also holds true for the First Nations, who have yet another way of cutting into the circle – by way of the impact of colonization and imperialism on aboriginal economies and of the resistance of aboriginal peoples to such encroachment and imposition of external systems of power, both capitalist and patriarchal. In short, once we admit the concept of patriarchy as having a dialectic of its own that cuts into and across that of capitalism, astounding untouched areas of study open up, areas that have been ignored simply because patriarchal thinking judged them to be either nonexistent or unimportant.

THE PROBLEM OF PRIVATE AND PUBLIC REALMS

The concept of patriarchy as a system of power establishes the legitimacy and primacy of "second" nature, thus cementing in ideology the legitimacy of the notion that some groups "naturally" possess power, the right to dominate and control the bodies (including labour, productive and reproductive) of those who remain tied to "first" nature. At the same time, power itself comes to be seen as a right possessed by the few. That is, unlike societies based in cooperative relations where power is not seen as either a right or a good because everyone is "empowered," "civilization" is marked by the redefinition of power as a value in itself, valuable because only the few can wield it (French 1985). Those who qualify for power are those who also can qualify for membership in "second" nature.

This splitting of human society into those tied to "first" and "second" nature must be enforced through physical coercion, precisely because such dichotomization is foreign to humans and people resist subjection. The emergence of a state responsible for enforcing the dichotomy and encoding it as part of the legal order is connected to the cementing of patriarchal power. Further, the establishment of patriarchy is also seen by those who wield power as the beginning of recorded history, of "civilization," in its modern sense.

In *The Creation of Patriarchy* (1986, 226), Lerner situates the beginning of history (as far as Western civilization is concerned) in the third millenium B.C., the point at which *men* (for women, according to Lerner, remained in pre-history, with a few exceptions, until the nineteenth century) began to record, define, and interpret the past. "Above all, human beings are concerned with immortality. The desire to survive one's own death has been the single most important force compelling humans to record the past and preserve it" (200).

Arendt, in her essay "The Concept of History" in *Between Past and Present* (1987, 41–90), argues that history as defined in the modern age is of recent origin. Marx's definition of history as process, "man making history" (an idea borrowed from Vico) mirrors the preoccupation of the modern age with conscious intervention in nature in order to bring about purposeful change. "The modern concept of process pervading history and nature alike separates the modern age from the past more profoundly than any other single idea" (Arendt 1987, 63). In early Greece, history had been viewed as remembrance of the *acts* of (specifically) men, and recorded to preserve their transcendence beyond the mortal life span of the individual to whom the act was credited. In fact, the individual has less importance than his act, the purpose of which becomes the drawing together of a community of men who separate themselves from their own mortality in communion with one another. "It [Homeric impartiality] could come to the foreground only after long experience in polis-life, which to an incredibly large extent consisted of citizens talking with one another. In this incessant talk the Greeks discovered that the world we have in common is usually regarded from an infinite number of different standpoints, to which correspond the most diverse points of view ... Greeks learned to *understand* – not to understand one another as individual persons, but to look upon the same world from one another's standpoint, to see the same in very different and frequently opposing aspects" (Arendt 1987, 51, emphasis in original).

According to Arendt the primacy of labour originated with the modern era – with capitalism. Before that time, labour had little or no public value and was hidden in the private realm. The origins of Western civilization were founded on consigning certain groups (especially women and slaves) to the private realm, to labouring for others. Their ties to necessity (to "first nature") established the basis (material, ideological, legal, cultural) for the "freedom" (from necessity, from labour, and from "first nature") of those who appropriated the labour of *women as a group*, and of slaves (Lerner argues that for long periods of time, perhaps centuries, slaves consisted of women and children, since captured warriors were generally killed; see also Chevillard and Leconte 1986 and in Coontz and Henderson 1986, 156–68).

In *The Human Condition* (1958), Arendt challenges Marx's labour theory of value; or, rather, she argues that this theory is applicable only to the modern era. She also rejects the dialectical method (especially its concern with process and the conceptualization of history as an unfolding of contradictory processes) as underlying the thinking of the modern era. To the emphases both on labour as at the core of human endeavour and on material dialectics as the method to understand the transformations wrought by "man" in and through nature, Arendt posits another way of being in the world: that of the Greek *polis*. In the course of her analysis, she presents a picture of the creation of the public and private realms as marking the beginning of Western civilization. I propose to use Arendt's political theory as an explanation of the acceptance, in Western philosophy, of men's conscious creation of Western civilization as a breaking into the cycle of nature by men, who then created their own history and politics as over and above the meeting of human needs per se. Civilization is then seen as somehow "natural" to men, in the sense of man's desire to overcome his own mortality. But in reality there is nothing natural about this. It represents a human and social creation of an entire power system that also necessitates a state apparatus to enforce it.

It is interesting to note that tracing the origins of patriarchy in Western civilization takes us back several hundreds of years *before* the modern definition of civilization came into common usage. That is, the system of power associated here with patriarchy had been in place for a long time before the modern idea of civilization adopted it. And although it was also in the nineteenth century that thinkers like Marx rejected patriarchal relations as outdated in the modern industrial era, the relations of power that the concept defines were then accepted as somehow natural. This can be seen, for example, in Marx's quote about the relation between men and women as founded in the "natural" power of men to subjugate women to their "communal lust" (Marx does not portray the sexual relation as in any way an equal meeting of the needs of both men and women). Marx then sees the development of relations of exchange through industrialization as the way by which such unequal relations of power will disappear.

Arendt's analysis in *The Human Condition* is pertinent because she distinguishes between labour and necessity as not human, on the one hand, and, on the other hand, the establishment of reason and the *polis* (a public realm of political action) as marking human life as different from other animal species, as breaking open the cycles of nature, including cyclical time (the marking of single great acts as history). For Arendt, then, the human condition of labour is life, the meeting of the necessities of existence. All humans are rooted in necessity, as are all animal species. But to be rooted in necessity is not to be human. Those who spend their lives in this way, the *animal laborans*, remain rooted in the world and in nature the same as all animal species, tied to the cyclical forces of nature.

The true insertion of "man" in the world is achieved only through action, the only activity between "men" not mediated through matter or nature. History, for Arendt, is the *remembrance* of these actions. And the *polis* is the area created by men where they can come together and act, as well as collectively remember the actions of others. Her examples of action are significant, since most, if not all of them, relate to warfare. The human condition of action is plurality.

It is interesting to compare Marx's concept of activity to Arendt's concept of action. Marx argues that all activity (which would include Arendt's more restricted notion of action) is rooted in subjectivity which is itself tied to labour. The ability of humans to cut into the cycle of nature is based in their ability to objectify what they do and to picture in their minds both their actions and the possible consequences of those actions. Marx would most likely call Arendt's definition of action a form of alienation. However, Arendt's point is equally valid to that of Marx. We can, in other words, accept Marx's insistence that all activity, including action in the public realm, ultimately stems from the consciousness of humans in and through collective labour.

At the same time, we can accept Arendt's point that the human condition itself comes to be defined as a *denial* of this basic necessity to labour, which is possible when the human condition is itself defined by those who, in a fundamental way, reject it. They can do this by categorizing the majority of people as not human, as tied by their natures to animal life, born in other words to labour and serve those who are qualified for a higher purpose. The labour that is then extracted from these people is first and primarily used in the production of the necessities of life for those who rule them, use value in the private realm of the household. As such, this labour in the household has no value. Household can now be defined as a social construction that solidifies patriarchal power which resides in those men who are recognized by the state as having rights to the territory encompassed by the household and to the people contained within its boundaries. These men acquire the rights of citizenship to a territory defined by a state (be it a city-state or a nation state).

Thus the history of Western civilization is founded on a particular set of relations that values neither labour in itself nor those identified with it. The Greeks created the public realm to insert "man" into the world through action which breaks the cycle of the *animal laborans*. However, Arendt notes that this division has been reversed in the modern age. That is, modern "man" has substituted the private for the public realm. Labour and necessity now become the primary aims of all humans, since the industrial *process* is based in relationships of exchange and the corresponding mode of production. Arendt laments that now activity, in the Marxian sense, and action, in her sense, have been merged. Action has now been channelled, so that "we have begun to act into nature, in the literal sense of the word" (Arendt 1958, 231).

Rather than situating themselves in a world which they have consciously erected to counter the timeless cycle of nature, in the modern age men have chained themselves to their own necessity, simultaneously drawing nature into their world. Arendt (1958, 134–5) views this as an extremely negative development, both for the way humans view themselves and for the potential destruction not only of the world they have created for themselves, but of all of nature, since the two have, in many ways, become synonymous.

In other words, the very forces Marx celebrates as leading to the liberation of humans from necessity and from the constraints of nature, Arendt argues enslaves "mankind" in nature, as an *animal laborans*, and offers the possibility of man's destroying nature. And in destroying the distinction between the private and the public (through the creation of "society," a concept Arendt notes was absent in Greek thought), those groups previously relegated to the private realm have lost all protection. For in being assigned to the private realm, they were guaranteed protection provided by those who ruled in the public realm.

According to Greek thought, the human capacity for political organization is not only different from but stands in direct opposition to that natural association whose center is the home (*oikia*) and the family. The rise of the city-state meant that man received "besides his private life a sort of second life, his *bios politikos*. Now every citizen belongs to two orders of existence; and there is a sharp distinction in his life between what is his own (*idion*) and what is communal (*koinon*)." It was not just an opinion or theory of Aristotle but a simple historical fact that the foundation of the *polis* was preceded by the destruction of all organized units resting on kinship. (Arendt 1958, 24)

The household was essential to the citizen. To be a citizen of the *polis* required freedom from necessity. This occurred in the private realm of the household, whose head was thus liberated for public life. "The realm of the *polis* ... was the sphere of freedom, and if there was a relationship between these two spheres, it was a matter of course that the mastering of the necessities of life in the household was the condition for freedom of the *polis*" (Arendt 1958, 30–1). Thus the household headed by the master-citizen was also a form of private property. In order to belong to the body politic, one had to have a sphere of one's own (61). And Greek law, *nomos*, from *nemein*, "to distribute, to possess (what has been distributed), and to dwell," (63, n.62) was originally confined to defining and enforcing the private space, by which the public was defined: "This wall-like law was sacred, but only the inclosure was political. Without it a public realm could no more exist than a piece of property without a fence to hedge it in; the one harbored and inclosed political life as the other sheltered and protected the biological process of the family" (64). "The organization of the *polis*, physically secured by the wall around the city and physiognomically guaranteed by its laws – lest the succeeeding generations change its identity beyond recognition – is a kind of organized remembrance" (198).

By defining humanity in terms of those who participate in *polis* life, those forced into the private realm are seen as animal-like, dependent on the good will (and laws) of humans. We can see here the degree to which women have been deprived of their humanity. They are naturally assumed to be tied to their biological natures, especially their functions of reproduction. They labour in nature to bring forth children who, if they are male, can be reborn into *polis* life. "Moreover, since action is the political activity par excellence, natality, and not mortality, may be the central category of political ... thought" (Arendt 1958, 9). O'Brien (1983) argues that there has been no Western philosophy of birth to match philosophies developed around other natural functions; for example, labour/Marx, sexuality/Freud, and death/existentialism/Sartre. However, as O'Brien herself notes, and as Arendt argues here, the whole tradition of Western philosophy itself revolves around "second" nature as a rebirth. In fact, the natural function of birth is elevated as the basis for Western thought, not in its biological manifestation, but as a renewal of life in thought (O'Brien 1983, 156–7).

Mills (1987 221, n.20) makes a further point that is pertinent here: "In *The Politics of Reproduction* Mary O'Brien claims that the redefinition of the proletariat from the original Latin meaning (the class that has no wealth but its children) to the meaning it has in Marx (the class that has no wealth but its labor) is significant. Capitalism, as a system that produces and reproduces itself, transforms the meaning of human *continuity* from that which occurs through human biological reproduction to that which occurs through economic relations" (emphasis in original).

By identifying history as beginning with socialism, and by accepting labour as the founding principle for human cooperation, Marx ignored the very historicity of the concept of labour in terms of how it is used to produce use values themselves. That is, as Arendt, Lerner, and O'Brien (each in her own way) have shown, production for local use, for immediate necessity (and this includes labouring to bring forth human life), in all patriarchal systems comes to be equated with what is not human, what is, in short, animal-like. This type of labour *loses* all value, and those who are forced to perform it (women, children, and slaves) are not judged to be human as long as they are tied to performance purely for the local use of those who control them.

This reformulation throws some light on the private/public dichotomization that has plagued much of socialist feminist analysis, the most extreme variant of which is the concept of the domestic mode of production (Armstrong and Armstrong 1983; P. Smith 1980; Zaretsky 1976). In a sense, much of this analysis also does not distinguish between the family and the individual (since women are automatically tied to the family and the two become conflated). A historical tracing of the roots of Western patriarchy in establishing a state and in consciousness itself suggests that kinship was the means underlying production for both use and exchange. The development of a class society whereby a group of people appropriated the surplus of the majority, not through relations of kinship but through political and religious domination, suggests that a historical understanding of the place of the family must be situated in the establishment of patriarchy. That is, women are not *necessarily* subjugated to their fathers and husbands. That they become oppressed in this way has its own historical roots in the ways in which men come to appropriate the labour of women and their children. The process of industrialization is then a *further* elaboration of this oppression, just as it serves to re-emphasize production for use, but with a new twist: production of use values that acquire value through exchange, through the creation of surplus that can be appropriated by the capitalist. The oppression is, in other words, not *given* or in any way *natural*. There is a danger of seeing it as such if Marxist analysts *begin* with the capitalist mode of production itself.

In Marxist analysis, the process of industrialization in the capitalist mode of production takes place when activities normally confined to household production are reorganized outside the home and are produced not for household use but as commodities sold for a profit. Thus activities previously performed in the private realm of the household become part of the public realm, undertaken for market exchange rather than domestic use. And because they are now organized around production in bulk, households can no longer afford to produce them as cheaply as factories can. Neither do labourers have time to keep up these private activities, as they are forced onto the labour market to sell their labour power. They thus buy back the commodities with their wages. The household is transformed from a unit of production and consumption to one of consumption.

Gender distinctions are then elaborated within this framework. The birth and rearing of children is confined to the household, as are women. Men continue to be linked to the public world, now elaborated along class lines of capitalist employer and proletarian labourer. When women enter the job market, they are segregated into jobs characterized as "women's work," and because women are unpaid when they work in the household, these jobs are very poorly paid. Much feminist thought has gone into trying to reconcile these sets of contradictions that revolve around women's unpaid work in the home, their lowly status in the paid labour force, and the notion that men earn the wage that provides their family with its basic necessities (most of which have been commodified). In the process, the private/public split tends to be assumed and recreated in feminist analysis.

But first, to dichotomize is to adopt the artificial distinctions that date back to the origins of Western civilization and patriarchy. Underlying all dichotomies is that of male and female, replicated here in public and private. Second, it is the establishment of the public realm (of a superimposed "second nature" of mind, soul, creativity, work upon body, matter, procreativity, labour, of "first nature") that *always* creates and defines the private. Both are *creations* of "man." This also suggests that Arendt's distinction between action and labour is not so absolute as she would have us believe, since the one is rooted in the other. The *belief* that action has nothing to do with labour is itself a patriarchal construction. Shiraz Dossa (1989, 49) reaches a similar conclusion: "Arendt isolates action as the peculiar quality of man and man alone. This capacity to act makes man a political and historical being, but man himself is not a 'product' of history. The capacity to act is not itself historical: it precedes history, it is ontological."

Third, as philosophers and political theorists spend much time demonstrating, the public realm of mind, philosophy, and politics has had a complex history in Western civilization and elsewhere. Thus the very nature and definition of the public realm has changed, through, to borrow Arendt's phraseology, the actions of men, their insertions in the world, with consequences that are unique, that reverberate on the acts of others, with resulting new configurations of what is meant by the public realm. As a consequence, the creation and definition of the private realm also changes, but not only because of the actions of those who rule (however this is defined), but also in and through the struggles of those whose history has gone unrecorded.

CLASS AND PATRIARCHY

To recognize that the conceptualization of labour expended in creating use values has also undergone transformation in the various historical periods requires us to disentangle the concept of patriarchy from that of class. Those feminist writers who have tended to focus on patriarchy, like Lerner (1986) have generally argued that class relations are part of patriarchal relations (Muszynski 1989). Lerner, for example, provides documentation tracing patriarchy and the foundations of Western civilization to the establishment of archaic states in Mesopotamia, Egypt (the Hebraic legacy), and Greece. She also traces the complex ways by which women were subordinated. But she provides a very confusing analysis of emerging class relations.

Thus at the outset Lerner provides a number of propositions. The first one is: "The appropriation by men of women's sexual and reproductive capacity occurred *prior* to the formation of private property and class society. Its commodification lies, in fact, at the foundation of private property" (Lerner 1986, 8; emphasis in original). This is the creation of patriarchy at the origin of Western civilization. But in her fifth proposition she argues that class is different for men and women. "Class for men was and is based on their relationship to the means of production: those who owned the means of production could dominate those who did not. For women, class is mediated through their sexual ties to a man, who then gives them access to material resources" (9). Lerner is in fact referring to two different historical phenomena here, and they cannot be collapsed the one into the other (Marxist analysts tend to collapse patriarchy into class, reversing the process). This reductionism leads her to conclude: "At any given moment in history, each 'class' is constituted of two distinct classes – men and women" (215). Having spent her entire analysis demonstrating the dichotomization that lies at the heart of patriarchy, in the end she reproduces the very dichotomy she seeks to destroy.

Perhaps one way to overcome this dualism while retaining both concepts is to search for an answer to the question: What is appropriated at any given historical moment? The concept of patriarchy allows us to answer the question by examining both the exclusion of *groups* of people from civilization and the establishment of categories as appropriate vehicles for thought – the categories, for example, of woman, of slave, of master, of race. The concept of class directs us to an examination of *economic* relations that refer to the appropriation of surplus from *recognized categories* of people. As Marx emphasized, these economic relations were themselves mediated in various ways (the intent here is not to reproduce a base/superstructure dichotomization). Labour cuts across both concepts.

For example, the Greek city-states were patriarchal and class based. Appropriation of surplus along class lines required recognition of certain *individuals*, who received recognition by and through the public realm. Heads of households, the citizens, were required to pay taxes to the ruler, and were thus exploited, since their surplus was appropriated.

The manner by which they in turn realized surplus created further class distinctions, depending on their relationship to the means of production. Shopkeepers, tradesmen, farmers, artisans were both differentiated in their economic pursuits and united as citizens, recognized as having certain ties to the public realm.

As heads of households, however, these individuals commanded resources that were human as well as material. And here we address patriarchal relations. Those confined to the private realm of the household were not recognized and were given no individual status. They were part of the means of production, whether they were the wives of the household head, his children, or his slaves. Class relations do not apply to these groups as long as they have no public voice. This does not deny the fact that class relations of exploitation were historically founded on group exclusion. Rather, it reinforces it by pointing to the *complexity* of the relationships thus established. It also points to the fact that not all humans were regarded as individuals. Our understanding of the term "individual" is itself the product of a long history, and has undergone a major transformation in the modern era (Lerner, in fact, treats all people as individuals in the modern sense of the term, ignoring its historical specificity).

Recognizing this distinction between the categorization of the group and the position of any one individual in relation to it throws some light on what may appear to be an anomaly. That is, while women as a group, for example, have been denied the recognition that individual men have been able to acquire, for instance, as citizens, this has not meant that women have never exercised power as individuals. Marilyn French (1985) documents numerous instances in the history of Western civilization when women exercised economic, religious, and political power. However, she also notes that these instances occurred when the hegemonic power structures were either in crisis or in a state of transformation (she gives as examples the establishment of Christianity, the Protestant Reformation, the French Revolution: cases where women engaged in struggles for change and exercised influence). She also notes, however, that the influence of individual women was generally short-lived. Once new power structures were in place, women were again subordinated according to the new ideology. This all points to the necessity for a coercive power apparatus to keep women (and other groups) subordinate. Given a chance, individuals will seek to act against the oppressive structures in place.

One set of complex relations that cuts across both patriarchy and class is contained in Arendt's and Lerner's discussions of *homo faber* as a creator of products fashioned out of his creativity and the materials available from nature. Arendt argues that such products acquire "use value," not in or of themselves, but because value is given to them elsewhere.

In so far as *homo faber* fabricates use objects, he not only produces them in the privacy of isolation but also for the privacy of usage, from which they emerge and appear in the public realm when they become commodities in the exchange market ... For it is only in the exchange market, where everything can be exchanged for something else, that all things, whether they are products of labor or work, consumer goods or use objects, necessary for the life of the body or the convenience of living or the life of the mind, become "values." This value consists solely in the esteem of the public realm where the things appear as commodities, and it is neither labor, nor work, nor capital, nor profit, nor material, which bestows such value upon an object, but only and exclusively the public realm where it appears to be esteemed, demanded or neglected. Value is the quality a thing can never possess in privacy but acquires automatically the moment it appears in public. (Arendt 1958, 163–4)

Thus Arendt argues that to characterize labour before the modern era as creating use value is nonsensical because neither labour nor its products was recognized in the public realm. The coming together of traders and of *homo faber* in the market to exchange the products of the latter's work created a public space, and thus specific works became valuable. However, as Marx himself argues, the centrality and primacy of commodification and of the market place was only possible when labour itself *acquired* value in the public realm, and became a commodity. And here we confront the confusion of the group with the individual. Marx notes that, in order to commodify labour, the individual as a possessor of labour power is necessary (and thus the redefinition of the public realm of the state as a liberal democracy). But historically, only certain groups could achieve recognition as individuals, and this fact does not disappear with industrialization.

There is an implied contradiction here between patriarchal relations of oppression and modern class exploitation within the capitalist mode of production, but it is only an apparent contradiction. It is a contradiction only if we acknowledge class relations as basic and primary within capitalism. In fact, patriarchal relations have had their own historical dynamic, not unrelated to the transformation of classes, but certainly not reducible to class. Further, there is a fundamental contradiction here that a focus on class relations hides; that is, the contradiction between the public recognition of the individual (historically limited to the few) and the public recognition of the possessor of labour power (the recognition of the majority of a population). This would imply that the state has recognized all labourers in a capitalist economy as individuals (indeed, the ideology of liberal democracy supports this view). However, public recognition of the individual continues to be denied to whole groups of people (women, immigrants classified into racial categories, aboriginal peoples, etc.). The old patriarchal exclusions continue, although they also undergo transformation to meet the demands of capitalists for cheap labour and in response to organized resistance by the excluded groups. For the public value acquired by labour within the capitalist mode of production gives visibility to groups who would otherwise remain invisible.

This analysis also suggests a way out from conflating women with the household, and here the present case study makes the distinction very visible. In the various commissions in British Columbia inquiring into the employment of various labour forces in such activities as fishing and shore work, salmon canners often stated that the ideal labour force for what they judged to be the less skilled work inside the canneries would have been (and in fact over time became) women. However, before the turn of the century, there was an imbalance of men over women in the non-aboriginal population. The labour power of women, and children, confined to the nuclear family household, was simply not available. Thus salmon canners and others turned to other *categories* of labourers (equally disenfranchised in the public realm) to provide this cheap labour. Chinese men were employed until they began to demand decent wages, at which point canners began to mechanize their lines. Immigrant women also served later, as did aboriginal women throughout the history of the canning industry. In each case, however, the group providing labour power was disenfranchised in a particular way. From this point of view, then, the labour power of dependent white women confined to the household is only one form of potential cheap labour power available to capitalist employers. And it is seldom the most devalued (or unvalued). As categories of people acquire rights as individuals (which primarily involve the rights of citizenship), their bargaining power as individuals and as a group is strengthened, although this does not necessarily mean that their labour power is priced in the same way as the most advantaged group: white skilled male labour power.

If class relations within the capitalist mode of production were in fact the only form of exploitation and oppression, a socialist revolution should have occurred as Marx predicted. Its postponement, indeed the increasingly bleak prospect of a revolutionary proletariat in the most advanced capitalist economies, indicates other forces at work. One of the most insidious and invisible is rooted in the transmission of patriarchal ideas through the ideology incorporated by and through the state and through individual socialization and enculturation.

CONSCIOUSNESS, IDEOLOGY, AND THE STATE

The concept of patriarchy has been used primarily in two senses here in order to throw some light on the transformation of use value in production. Lerner, for example, treats it as a historical phenomenon, arguing that the process of its development can be traced back in time. However, there is perhaps an even more profound way of approaching patriarchy, using Marx's approach to the implications of the primacy of exchange value in transforming production for use.

In his writing, Marx notes that exchange value exists alongside production for use in many different modes of production and in many different periods of time and areas of the world. It is only with capitalism, however, that the full implications of production primarily for exchange emerge and thereby transform all other relations, including the way in which humans think about these relations – consciousness. "Consciousness is seen from the beginning as part of the human material social process, and its products in 'ideas' are then as much part of this process as material products themselves. This, centrally, was the thrust of Marx's whole argument" (Williams 1978, 59–60).

And the production of "ideas" leads us to yet another problematic concept – that of ideology. Williams notes that the term was coined in the late eighteenth century by the philosopher Destutt de Tracy to mean the "science of ideas," thus locating it squarely within the empiricist tradition. But in Marxist writing three versions of the term have been used, often in contradictory ways: ideology as a "system of beliefs characteristic of a particular class or group"; as a "system of illusory beliefs – false ideas or false consciousness – which can be contrasted with true or scientific knowledge"; and/or as "the general process of the production of meanings and ideas" (Williams 1978, 55). Williams notes that the third version undercuts the other two, allowing for both a general process and its specific manifestations, particularly in class-based societies.

This allows us to introduce the concept of patriarchy within the *broader* process, as a formation of the consciousness of "man" transmitted through the socialization process almost unthinkingly. The work of Freud on civilization is interesting in this context because Freud argues, in *Civilization and its Discontents* (1982), that the establishment of "civilization" requires a certain consciousness that represses natural instincts, including the instinct for pleasure. Freud links the founding of Western civilization with the acceptance of the Oedipal complex as a way of organizing men in social groups: the acceptance of the primordial "fact" of the son's killing his father and committing incest with his mother (these two acts are somehow, for Freud, located at the very centre of the son's desires and in his choice of objects to satisfy these desires).

If we adopt the type of analysis undertaken by Mitchell (1975) and Rubin (1975), we can question what Freud claimed to be the primeval condition and see it instead as the foundation for an ideology rooted in patriarchy; in other words, the justification for patriarchal consciousness. What Freud was then describing was the process of socialization of infant boys and girls so that they would take their place within a patriarchal society without questioning their assigned roles. Freud can be accused of accepting this condition as the price "man" had to pay for becoming civilized. At the same time, however, he was aware that the price paid was extremely high, and the penalty was the development of all sorts of neuroses and psychoses. His own views on the "natural" place of women in his analysis are contradictory; for example, while he borrowed and adopted phrases like "anatomy is destiny" to explain the inferiority of women, he held individual women psychoanalysts, including his own daughter Anna, in high regard (Gay 1988, 501–22).

For "consciousness and its products" are always, though in variable forms, parts of the material social process itself: whether as what Marx called the necessary element of "imagination" in the labour process; or as the necessary conditions of associated labour, in language and in practical ideas of relationship; or, which is so often and significantly forgotten, in the real processes – all of them physical and material, most of them manifestly so – which are masked and idealized as "consciousness and its products" but which, when seen without illusions, are themselves necessarily social material activities. (Williams 1978, 61–2)

Patriarchy, then, can be understood as underlying a certain type of consciousness that dichotomizes the world into categories. Dorothy Smith (1987) uses the term "bifurcated consciousness." This is not simply a thought process but is reflected in the categorizing that excludes the vast majority of people from the human condition in order to appropriate their labour for the maintenance of that small group that sets itself up as superior, as human, as "man." That this is a social construction, and not biologically "given" in the human condition (rejecting the view that some categories of humans are superior to others *because* of their biology), is proved by the necessity that those who rule in this way must erect a state apparatus to enforce their domination. It is not enough that most infants born into patriarchal societies are trained from birth to know their proper place in the world, to accept their inferiority if they are born into the wrong group. In spite of patriarchal consciousness, humans still continue to resist their assigned places, and one of the primary roles of the state is the creation and recreation of the private/public spheres in order to reflect the historical reality at any one "moment."

The capitalist mode of production requires just such a redrawing of the boundaries between private and public, largely because, as we have seen, labour is no longer confined to the private sphere of the household. That labour becomes part of the now public sphere of the economy, however, does not mean that the ideological role of patriarchy becomes redundant. On the contrary, as noted earlier, it is very interesting that writers like Freud and Arendt have re-evaluated in a positive light the modern version of the concept of civilization, which assumes patriarchal consciousness as a given condition of humanity. Marx was also guilty of such an assumption.

In fact, in his discussion of the continuing attraction to modern "man" of Greek mythology and art, in "Introduction to a Critique of Political Economy" (Marx and Engels 1959), Marx characterizes the ancient Greeks as "children," an analogy Freud also uses to characterize early human history. For Freud, and perhaps also for Marx, this history becomes part of the consciousness that marks civilized society and that is subconsciously passed down through the generations. The sexual division of labour, and the domination of women by men, is then seen as part of the earliest infancy of humanity, that upon which all the rest has been erected. Althusser (1971) also has an interesting discussion of these issues in his two essays entitled "Freud and Lacan" and "Ideology and Ideological State Apparatuses (Notes towards an investigation)."

This beginning is as animal as social life itself at this stage. It is mere herd-consciousness, and at this point man is only distinguished from sheep by the fact that with him consciousness takes the place of instinct or that his instinct is a conscious one. This sheep-like or tribal consciousness receives its further development and extension through increased productivity, the increase of needs, and, what is fundamental to both of these, the increase of population. With these there develops the division of labour, which was originally nothing but the division of labour in the sexual act, then that division of labour which develops spontaneously or "naturally" by virtue of natural predisposition (e.g., physical strength), needs, accidents, etc. etc. Division of labour only becomes truly such from the moment when a division of material and mental labour appears. (The first form of ideologists, *priests*, is concurrent.) From this moment onwards consciousness *can* really flatter itself that it is something other than consciousness of existing practice, that it *really* represents something without representing something real; from now on consciousness is in a position to emancipate itself from the world and to proceed to the formation of "pure" theory, theology, philosophy, ethics, etc. (Marx 1988, 51–2; emphasis in original)

Using the concept of patriarchy we can argue that consciousness as ideology in the manner Marx develops it in this quote underlies his thinking from the beginning of his argument. That is, to see certain human activities, including sexual activity between men and women, as "sheep-like" already suggests a dismissal of social formations (particularly as developed in production centered around kinship – social formations that anthropologists and archaeologists are showing to have encompassed vast periods of time and huge territories) that were central to human history.

However, Marx did recognize that the formation of the state involves the dissolution of the more ancient communal ties in favour of the interests of the individual. "And out of this very contradiction between the interest of the individual and that of the community the latter takes an independent form as the *State*, divorced from the real interests of individual and community, and at the same time as an illusory communal life, always based, however, on the real ties existing in every family and tribal conglomeration ... and especially ... on the classes, already determined by the division of labour, which in every such mass of men separate out, and of which one dominates all the others" (Marx 1988, 53–4, marginal note in original with emphasis as shown). Unfortunately, Marx's analysis of the place of the state in his thought was left for a later project which he was unable to complete.

DEFINING THE CONCEPT OF PATRIARCHY

In focusing on class relations, especially their maturation in the capitalist mode of production, Marxist analysts (including Marx himself) tend to ignore the earlier history, and take for granted developments that most likely took much longer than the period that covered what is accepted as the recorded history of Western civilization. The establishment of Western civilization is synonymous with the establishment of a certain way of thinking about the world of men, which thereafter is taken for granted. However, this does not mean that patriarchal thinking is static. As class relations become the visible signs of inequality, patriarchal relations undergo transformation and also serve to define class. The two concepts must be analysed in a dynamic (dialectical) fashion.

Also implied here is the need to rethink certain of the categories that inform the concept of class. Labour is one such element, as already noted. Division of labour is another. While capitalists utilize labour power in a unique division of tasks, older divisions based on criteria such as gender and age can no longer be taken as universally given. They too are historically specific, not based on immutable biological characteristics. Implied here is a critique not only of Marxist analysts, but also of those radical feminists who argue that because of their biology, women share mothering or nurturance (implicit in which is a certain universalistic definition of motherhood).

Leibowitz (1986) challenges the notion of a sexual division of labour as physiologically necessary by arguing that hominid populations had a very brief life span. Thus, the bulk of the group consisted of a large number of children, and she argues that they in fact did a large part of the food acquisition for the group. Since marked sexual differences in terms of size and strength occur during puberty, it is unlikely that food acquisition was based on gender differences (and she presents archaeological evidence to support her argument). Thus she argues that the gender division of labour is itself a historical product, connected to increased size of the population and increased longevity, as well as the development of the forces of production (for example, the use of fire to cook food).

Two other important elements in the concept of class are the notion of private property and the rise of a civil state. The concept of patriarchy can be used to make sense of the emergence of these two elements by exploring the development of the public and private realms (necessary to the establishment of private property relations and a civil state). Particularly important here is the exploration of how establishment of private property relations enforced through a civil state served to destroy older kinship ties and to redefine the meaning of kinship through the state and related to the emergence of new ruling classes (Coontz and Henderson 1986; Lerner 1986).

Finally, the concept of patriarchy requires that we address the construction of the individual in particular cultures and periods, particularly how the idea of individuality is rooted in the categorization of the groups excluded. For example, Coontz and Henderson (1986) note that the emergence of lineage societies was marked by the emergence of property owned by the kin group. This type of ownership was related, in turn, to patrilocal residence upon marriage and the appropriation of women as property to be exchanged between groups of male kin. We see here the beginnings of the structured domination of women by men. Coontz and Henderson also note that not all lineage societies displayed such domination, and it is important to explore the differences among them (one important factor in the subordination of women is patrilocal residence, since women are separated from their kin groups).

In the emergence of civilization, it appears that gender was the defining characteristic in the subordination of women. That is, once the category "woman" is constructed and tied to first nature, exceptions can be made as to who fills specific positions. For example, in discussing the practice of slavery in certain African lineage societies, Chevillard and Leconte (1986, 157) note that male captives usually acquire female status: "He will not be entitled to a wife, will not have progeny, and in practice, will be required to perform 'women's work.' If he is given a wife after some time, he will then lose his captive status ... Socially speaking he is merely a 'sterile woman.'" This suggests that there is a category of woman that carries a certain devalued status to which individuals are allocated, a status not necessarily based on physiological characteristics. In particular, these authors, as well as Lerner (1986), suggest a connection between the category of woman and that of slave. The two are by no means identical, but they do have historical commonalities as well as overlaps (for example, the argument that most early slaves in archaic societies were women). This suggests that the concept of race may also be fruitfully explored by using the concept of patriarchy.

For example, in the last half of the nineteenth century there was an imbalance between white men and women (and children) in what became the province of British Columbia. In an industrializing economy, certain jobs had to be filled. These had been typed as "women's work" in western Europe, but few white women were available to fill them. Chinese men were employed in their place, notably as domestic servants, in the salmon canneries, and in laundries. In the royal commissions held in 1885 and 1902, provincial politicians and employers referred to the Chinese as a "feminine race." Rather than acknowledge that these were men, those with political and economic power conflated their race with the gender of women.

The reverse can also occur, but only on an individual basis. That is, once women as a group are subordinated to men as a group, individual women can fill vacant positions generally reserved for men. Their being allowed to do so is predicated on the fact that most women are oppressed, and can in fact serve to render that oppression more invisible through tokenism. More often, individual women fill positions of power when male kin are unavailable (and it is generally acknowledged that these are exceptional circumstances). These women, then, become men, at least as long as they are occupying these particular positions. I think such anomalies are better explained by using the concept of patriarchy than that of class.

CAPITALISM: THE CASE OF THE B.C. FISHERIES

In terms of examining the creation of a cheap wage labour force in the B.C. fisheries, it is important to link an understanding of class relations to the categorization of patriarchal thinking as it emerged within Western civilization. This allows us to understand the differential proletarianization of different *groups* engaged in shore work and in fishing (Chinese men, aboriginal women, aboriginal men, Japanese men, Japanese women, and white men in the early phase). Class, race, and gender thus become equally important and problematic, as well as intertwined in the struggles that ensued among various groups (again differentiated along lines of class, race, and gender, and further complicated by changing alliances along all these dimensions at different times and in different places along the B.C. coast).

The focus on the emergence of patriarchy in Western civilization is central in this chapter because it was implicated in the emergence of capitalism. And confining the categorization of gender and race to Western civilization forces us to treat categorizations that develop elsewhere in a nonreductionist manner. That is, we cannot assume a certain evolutionism according to stages of economic, or any other type, of development. While all civilizations are patriarchal, historical expression of patriarchy differs in each.

It is therefore necessary to study how patriarchy within Chinese civilization structured groups and classes. In other words, it is not sufficient to merely argue that European imperialism recreated its own patriarchal relations abroad. The release of Chinese male peasants and their migration to the Pacific Northwest was connected to European imperial expansion and to internal developments of patriarchal and class relations(the extreme devaluation of women and the impoverishment of the peasantry, the last aggravated by European invasions). And the employment of Chinese male peasants in the nascent industries of the Pacific Northwest reflected the European thrust to industrialize North America as well as a negotiation for the employment of certain groups (involving elements of class and patriarchy, both European and Chinese).

To complicate matters further, we cannot assume that the aboriginal population of the Pacific Northwest was still, at the time of European conquest, living in a communistic economy (if it ever had one). Knight (1978, 26) contends that "prevalent popular views generally disregard or gloss over considerable evidence of suffering, hardships, and oppression between and within indigenous Indian societies." Evidence has been presented that dates the Prince Rupert (Coast Tsimshian) area back five thousand years, and shows human occupation on the Columbia River for ten thousand years (Suttles 1987, 270). Thus the history of Pacific Northwest coast peoples is older than that of Western civilization. We cannot assume that throughout this period they pursued a simple and static hunting/gathering/fishing mode of production. There is great linguistic and cultural diversity within the area, and at the time of conquest a stratification system existed which included slavery and a well-defined gender division of labour (for example, throughout the area processing fish catches and preserving them for winter consumption was "women's work").

It is beyond the scope of the present work to delve into the establishment and development of patriarchy in China (and Japan) or to explore whether women as a group were dominated by men as a group in the various indigenous economies comprising the Pacific Northwest. Instead, in the following chapters, some of the dynamics outlined here will be explored in the specific context of the establishment of a fishing industry in British Columbia.

3 The First Nations, Property Rights, and Salmon Canning

THE EFFECTS OF PATRIARCHY AND COLONIALISM ON THE CONSTRUCTION OF RACE: DEFINING THE CATEGORY "INDIAN"

The preceding chapter treats the gendered division of labour as part of the problematic defined and discussed in the introduction. If the work that men and women do is valued in different ways, then the labour theory of value must take this factor into consideration. And if, as has been argued, the labour of women is considered to have no value, then the participation of men and women in the paid labour force within the capitalist mode of production cannot be treated according to a uniform and abstract set of laws. That is, the participation of women is already conditioned by the participation of men. The position of women has already been subsumed to that of men who hold power at many different levels in a patriarchal society: as fathers and husbands, employers, law makers and enforcers, etc.

In *The Second Sex* (1989, 65), Simone de Beauvoir refers to woman as "Other" in relation to man: "Men have presumed to create a feminine domain – the kingdom of life, of immanence – only in order to lock up women therein." Dorothy Smith refers to bifurcated consciousness; a method of thinking that treats man and woman as oppositional categories and that further extends this method of dichotomization to all aspects of life and nature. Opposites in this sense are not equal but rather the primary term is established as superior while the opposite is just that, inferior by virtue of being the opposite to that which names itself as primary. "Women's means to reflect upon themselves is a reflection from outside themselves, the structuring of themselves not as subjects, but as other. Furthermore, in its contemporary terms, it appears not as men's view of women, but in impersonal and general terms" (D. Smith 1987, 52).

Kessler and McKenna (1985, 4–5), working within an ethnomethodological perspective, have yet another name for what is essentially the same phenomenon. They call it the "natural attitude." "In the natural attitude, there is reality and constancy to qualities like race, age, social class, and, of course, gender, which exist independently of any particular example of the quality. It is a fact that there are two genders; each person is a mere example of one of them; and the task of the scientist is to describe, as accurately as possible, the constant characteristics that define male and female, for all people and for all time. This is reality in Western society" (5). At the end of their work, the authors evaluate the implications of this social construction.

> In the process of attributing "male" or "female," dichotomous physical differences are constructed, and once a physical dichotomy has been constructed it is almost impossible to eliminate sociological and psychological dichotomies. Given that the physical dichotomy is androcentric, it is inevitable that the social one is also.
>
> Whenever science has offered evidence of a biological continuum, but everyday members insist (because of the way reality is constructed) that there are discrete categories, there have been attempts to legislate against the continuum. Laws in the United States on what constituted a "Negro" and laws in Nazi Germany on what constituted a Jew are two of the most obvious examples. These laws did not reject biology, since biology is a crucial part of the construction of Western reality, but used biology. (Kessler and McKenna 1985, 164)

Another "obvious example" is the Canadian legislation embodied in the Indian Act which defines what constitutes an "Indian" and breaks the category into sub-categories like status and non-status Indians. Whatever the implications for actual individuals, the fact is that an entire group is not only seen according to the category thus legislated but also comes to see itself as meeting or not meeting the criteria that constitute "Indianness" as enforced by the federal department established to carry out the terms of the Indian Act. For many of Canada's First Nations this has led to inter-group tension and conflict. A prime example of this is the difficulties resulting from Bill C-31 (1985) which amended clause 12(1)(b) of the Indian Act to re-incorporate status Indian women who had married non-status men and had thereby been deprived of Indian status for both themselves and offspring of the union. While the change in the law rights a wrong for the patriarchal treatment of Indian women whose status was defined by their marriage partner, it places hardships on over-crowded reserves and bands which have not received any additional financial or material compensation to re-incorporate large numbers of individuals who are now "Indians" once again. The authors of *Canadian Women: A History* (Prentice, Bourne, et al. 1988, 396–8) also recount the organizational efforts of aboriginal women, going back as early as 1968, to right this wrong, requiring recourse to the United Nations Human Rights Committee which ruled in 1981 that the Canadian government had contravened the International Covenant on Civil and Political Rights.

The legal definition of "Indian" also becomes a major factor in legislating aboriginal claims to land and to fishing sites and thus comes to constitute a reality in its own right, as existing in a "law" which the First Nations had no part in constructing in the first place. But the law then comes to define who they are as categories of people. It has also been the basis for denying them basic rights of citizenship since "status Indians" were "wards" of the federal government, much as women and children once were legal wards of their husbands and fathers. While the two realities were very different, the basis for the construction of categories like "Indian," "woman," and "child" were founded upon similar principles in British law and reconstructed in the Canadian legal system.

A final point to be made in this connection concerns the implications for aboriginal peoples introduced in 1982 with the new Canadian constitution. New categories have become legal realities. Paul Tennant (1990, xi) notes: "'Aboriginal' is now permanently enshrined in Canadian usage through the Constitution, which recognizes the Indian, Métis, and Inuit peoples as the 'aboriginal peoples' of Canada. 'Aboriginal' does not mean first on the scene; it means, rather, present when modern history began or when colonizers arrived. Indians were and are the aboriginal peoples of British Columbia." The Constitution included a Charter of Rights and Freedoms. Unlike the Indian Act, this was legislation over which the First Nations insisted they had a say. Berger (1982) provides the details. On 15 December 1980 the Nishga (Nisga'a) "appeared before the Special Joint Committee of the Senate and the House of Commons on the Constitution," and on 29 January 1981 "the Joint Committee agreed to recommend an amendment to the new Constitution, which provides: 'The aboriginal rights and treaty rights of the aboriginal peoples of Canada are hereby recognized and confirmed.'" While the prime minister and premiers decided to delete this provision, they were forced to restore it "by an outraged public opinion" (249). However, this was done "grudgingly, and in a qualified fashion. Now it is 'existing' aboriginal and treaty rights that are recognized and confirmed. None of the other measures in the Constitution and the Charter is limited in this way." However, Mr Justice Berger concludes that the wording is binding on the provinces as well as on the federal government (250). Since the province of British Columbia had previously denied any acknowledgment of aboriginal rights or claims, this provision was of special importance to the First Nations of that province.

In this study I use the term "First Nations" for two reasons. First, it is a term coined by aboriginal peoples as referring to themselves rather than a term imposed upon them, like "Indian" (although many aboriginal peoples refer to themselves as Indians) (Tennant 1990, xi). Second, the term implies that aboriginal peoples had political, economic, cultural, social systems that were at least equal to those of the Europeans. Although they had no formal state, since state develops along with class relations and a formalized hierarchical system of ruling like the feudal monarchies and capitalism, they had what Tennant (1990, 8) refers to as a "political tradition": "The higher population densities and smaller territories of the coastal peoples made specialized politics unavoidable." The First Nations were not nation states, but they were cohesive, dynamic, and self-contained entities with extensive systems of cooperation, trading, and conflict among one another. The colonial experience transformed these relations in profound ways without, however, destroying them.

In examining the further categorization of women, children, Indians, etc. as cheap wage labour, it is essential to take into account not only the establishment of patriarchal relations but also the period of colonization in what became British Columbia and the imperial expansion of Great Britain to China and Japan, resulting in economic difficulties for certain areas of these countries and leading to the creation of groups of migrants looking to improve conditions for themselves and their families by searching for work in North America.

Prior to capitalist exploitation and underlying it is a system of oppression. Paulo Freire, in *Pedagogy of the Oppressed*, defines oppression as the initiation of a process of violence in the sense that it denies humanity to the oppressed. "For the oppressors, 'human beings' refers only to themselves; other people are 'things.' For the oppressors, there exists only one right: their right to live in peace, over against the right, not always even recognized, but simply conceded, of the oppressed to survival. And they make this concession only because the existence of the oppressed is necessary to their own existence" (Freire 1990, 43).

Freire's argument connects to those developed here that women and children were denied their humanity in precisely this way with the establishment of patriarchy. In order to be fully human one must join "mankind" as a civilized being possessing the ability to reason. This is both learned (by male children) and an inherent property of men and not of women (first and second natures). But colonization and the imperial era further extended the notion of first and second nature through a process of classification developed by the European colonizer. Civilization became a European phenomenon, "the white man's burden," and it became the mission of the "white man" to take civilization to the "savages" and "barbarians" brought under foreign domination and rule. Miles (1989, 20) notes: "And when colonisation became an objective, a class of Europeans began a new era of contact and interrelationship with indigenous populations, a contact that was increasingly structured by competition for land, the introduction of private property rights, the demand for a labour force, and the perceived obligation of conversion to Christianity. Collectively, these were all embodied in the discourse of 'civilisation.'"

According to Miles, this discourse was developed within a changing reference system of "self" and "other." While the history of the European nation state rests on a discourse of who is to be included within a community by virtue of who is not included, inclusion/exclusion did not initially receive a positive/negative value. The Other could be different but not necessarily inferior. Categorizing the Other as inferior developed within the history of colonization, and incorporated the development of a changing ideology that we label *racism*. Miles (1989, 3) argues that the "way in which racism 'works' has certain similarities with other discourses," and he specifically names sexism. Simone de Beauvoir (1989, xxix) makes a similar point: "For whether it is a race, a caste, a class, or a sex that is reduced to a position of inferiority, the methods of justification are the same. 'The eternal feminine' corresponds to 'the black soul' and to 'the Jewish character.'"

Returning to Freire's argument, then, the concept of race is developed to deny the Other personhood. Humanity becomes a thing, an inherited property that is biologically determined; it is innate and unchanging and distinguishes between human and animal. For Freire (1990, 48) peasants are seen as being "almost submerged in nature." Referring to the work of Alvaro Vieira Pinto, he argues that consciousness itself becomes the object of a concerted method (56). That is, the oppressor is interested in changing the consciousness of the oppressed in order to more easily dominate them. Oppression comes to be experienced as "overwhelming control." In his essay in *The Politics of Education*, "Education, Liberation and the Church" (1985, 139–40), Freire also notes that the concept of the "Third World" is ideological and political, not geographic: "The so-called First World has within it and against it its own Third World ... The Third World is in the last analysis the world of silence."

Freire's discourse fits within that of Miles, which in turn fits within the discourse of patriarchy discussed in previous chapters. Indeed, the discussion can be further elaborated to include the oppression of aboriginal peoples. "The Okanagan and other Indians in Canada who fall under the Indian Act and live within the economic, territorial and political confines of reserves are not wards of the federal government, as is sometimes said, but members of little colonies within the borders of the controlling nation" (Carstens 1991, 285). Thus, the so-called "First World" also contains within its territorial boundaries its own "Third World," while at the same time it has also named whole spheres of the globe as "Third World" countries. Furthermore, Freire's conceptualization of "the world of silence" helps to explain how labour forces, like shoreworkers in the B.C. fisheries, were created in this world of silence, and why they were, initially at least, invisible relative to other labour forces, like fishers.

In her 1994 study of the European witch hunts, Anne Llewellyn Barstow also draws connections between the degraded status of women under patriarchy in the centuries from the Renaissance through the Enlightenment and the colonial period initiated by the Spanish and Portuguese, devoting an entire chapter to these interconnections.

> Having persecuted the Moslem and Jewish Other back home, the Spanish did not hesitate to do the same to the Americans. One Spaniard observed that they enslaved so many that the Indians "no longer approach their wives, in order not to beget slaves." Another claimed that "the Spanish treat them worse than slaves." The Spanish philosopher Sepulveda called the Mexicans "uncivilized, barbarians, wild beasts." He lumped together Indians, women, and animals, all creatures without a soul of their own. Considered unable to govern themselves, they were identified with savagery and evil. (Barstow 1994, 161–2)

Miles notes that in the second half of the nineteenth century, a merchant and industrial bourgeoisie emerged which began to seek political self-determination within the capitalist mode of production. He also situates the "ideology of nationalism" within the discourses of racism and sexism: "To this end, an ideology of nationalism was articulated which partially transformed the myth of the biological superiority of the Anglo-Saxon 'race' in order to create a sense of imagined community that was spatially and culturally distinct from its parent" (Miles 1989, 90). While he is referring to Australia, the case can equally be made for the emergence of a "white" Canada.

As illustration, the debate around the disruptive effects of the fur trade on the aboriginal population is instructive. While Fisher (1977) makes the case that the fur trade allowed a certain autonomy to aboriginal peoples and thus did not significantly disrupt their culture, Carstens (1991, 30) argues the opposite in regard to the Okanagan peoples. "It is true that there was an absence of major conflict in that the fur traders did not necessarily report such conflicts as major; but the fur trade did in fact radically alter both the internal nature of Okanagan society as well as Okanagan relations with the outside world. Indeed, the fur trade began the process which made the Okanagan subservient to the white man not just economically, but politically as well."

The core of Carstens' thesis revolves around the argument that the fur trade and white leaders who have come to be seen as "benevolent" in their treatment of aboriginal peoples (like James Douglas, chief factor of the Hudson's Bay Company's base on Vancouver Island and subsequently governor of British Columbia, and Gilbert Sproat, chair of the commission established in 1875 to investigate the "Indian land question") actually prepared the groundwork for the period of coercion that coincided with the settlement period.

> This period, lasting roughly half a century from 1811 to the early 1860s, produced structural and cultural changes which could never be reversed. In addition, radical and overwhelming changes took place in the personalities of individual Okanagan, and in the conscience collective of their society itself. The crux of these less obvious changes was the perceptions that the people developed of themselves. Whether they wished to accept it or not, they fell under the hegemonic spell of the white man and his institutions ... In short, they began to experience a dual or bifurcated culture and developed social personalities to match. (Carstens 1990, 52)

It was also in the period referred to by Miles, in the second half of the nineteenth century, that a merchant and industrial bourgeoisie emerged in what became, in 1871, the province of British Columbia. The merchant class, initially centred in Victoria, was central to obtaining the financing and securing the supplies and markets for the industrialists, the salmon canners who began to industrialize the fisheries in earnest in the 1870s. As Newell (1993) argues, the First Nations, or the "Indians" as defined by the Indian Act, were crucial to the prosecution of the industry, because of a need for both their labour in fishing and processing and their knowledge of the coast fisheries (which were central to the aboriginal economies of the coastal groups). However, my argument is that the consciousness of canners and those who held power in Victoria and Ottawa was a central factor in how they "saw" these labour forces. Further, the perception of "Indians" as "Other" was in turn mediated by the actual material conditions of all of the groups concerned. Salmon canners could only make use of "Indian" labour as cheap wage labour because of the ability of the latter to continue to survive in both their own world and that of the white man. The First Nations continue to struggle over land claims and food fisheries because their rights were not extinguished and because, more important, they would not allow these rights to be extinguished despite their erosion by the expansion of settlement and the establishment of capitalism by the white man. And the perception that capitalism was under the control of the "white man" was also part of the "conscience collective." Indeed, as will be shown, the fact that the Chinese in North America had access to considerable financial resources in China and began to invest them in their own factories (for example, making cigars in San Francisco), thus becoming capitalists, was considered to pose a major threat for the development of a "white" British Columbia and resulted in legislated exclusion of Chinese immigration to Canada.

TRANSFORMATION OF THE ABORIGINAL MODE OF PRODUCTION

To situate a study of the fisheries of British Columbia with the beginning of industrialization, of the introduction of the capitalist mode of production, is already to cut into history in such a way that what came before is necessarily lost from view except in how it affects capitalist expansion. Thus the thousands of years that encompass the history of the First Nations of the Pacific Northwest coast become invisible. Focusing on the process of industrialization also seems to bear out Marx's point, so hotly disputed in chapter 2, that history only truly begins with the end of capitalism and that capitalism is in this sense the pre-historical period that leads to its own demise.

This conundrum bears out Laslett's (1976) view of history as a "theory-laden activity." It also allows for recognition of the scholarship undertaken in many fields that explores the history and civilizations of the First Nations. Such a history, as just noted, is outside the scope of this study, but it is imperative to emphasize its "reality." As far as my study is concerned, what needs to be noted is the fact that prior to European contact, the First Nations occupied most, if not all, the territory that became British Columbia. Prior to European contact, the First Nations formed a fairly dense concentration of diverse cultures and social groupings. Duff (1977, 8) notes: "Except for barren and inaccessible areas which are not utilized even today, every part of the Province was formerly within the owned and recognized territory of one or other of the Indian tribes." Tennant (1990, 3) concurs: "The northwest coast of North America was a place of peoples. In the portion that would become coastal British Columbia, along the contorted shoreline, on the countless islands, and far inland along the many rivers, there was an extensive maritime population. Nowhere north of Mexico were there greater numbers or denser concentrations." Contact with Europeans decimated the population due to the spread of disease. Newell (1993, 21) estimates that the provincial population dropped by 80 percent between 1774 and 1874. And Tennant (1990, 3) notes that estimates for the precontact population have been revised to between three and four hundred thousand for the province as a whole.

While the population as a whole was large, the "building blocks out of which the tribes or bands were constructed were the small localized groups of people who lived together throughout the year." Membership followed "local rules of kinship," but the rules varied widely from one group to another. "These were the groups that owned the resource areas (though concepts of ownership also varied widely)" (Duff 1977, 16). In general, stratification increased as one travelled from the Fraser River north to Alaska. Tennant also distinguishes between interior groups like the Okanagan and coastal groups who were most directly involved in the fisheries that capitalists entered, beginning with the salmon canners. He notes that the coastal groups, according to anthropological studies, fit the type that "is highly structured and has distinct politics based on rank, status, and hierarchy. This type emerges only among substantial populations, which in turn require bountiful natural environments. The other type is less structured and much more egalitarian" (Tennant 1990, 6).

The First Nations had no political institutions of the nature of a state or a kingdom. Suttles notes (1987, 229): "Aboriginally, the village had its clearest identity vis-à-vis other villages at the potlatch. But this was the time of all times when individuals or kin groups acted separately to establish and maintain status, both in relation to other villagers and within the larger social unit of intervillage relations. The village had no overriding importance." The potlatch was an overarching economic institution which served to redistribute food and wealth throughout the area, and prevented any one group from amassing too much wealth, and, consequently, power (Suttles, 1960). This in turn reflected the material conditions of the entire coastal area and the rights of certain groups to have access to and to use certain sites, like the diverse fisheries. The food they obtained was distributed in the potlatch, and the rights to the areas and resources from which the food was taken were consequently reaffirmed by the groups who attended the ceremonies. Suttles (1987, 229) notes: "In the past the major intergroup gathering, the potlatch, was the most spectacular part of an economic system by which the Coast Salish population survived in a natural environment of fluctuating productivity." The five species of salmon, for example, run on multi-year cycles that differ according to species and area of spawning. Thus, a group having access to a major sockeye-producing stream on the Fraser River could have far more fish than needed for the village in one or two years only to be faced with starvation in the other two years. The potlatch helped to redistribute food supplies in an elaborate system of exchange that incorporated the entire province and that allowed individual groups access to resources they would otherwise not have, including food. Thus the population as a whole enjoyed access to a diversity of foods (among fishes, the various species of salmon, eulachon, herring, roe, halibut, etc.), and the potlatch also acted as an insurance against starvation since those without food one year attended the ceremonies and thus validated the claims of their hosts who would then reciprocate another year. It was this system that the European colonizer cut into and truncated. It is significant that one of the first aboriginal institutions Canadian law outlawed in the Pacific Northwest was the potlatch.

In summary, the coastal tribes had a wide variety of social groupings around comparable economic units based on variously defined kin associations. The total economic system consisted of a large number of loosely associated units, some internally stratified by rank, each claiming rights to certain key fishing, hunting, and berry and root gathering sites. This particular economic organization appears to have made ideal use of the environment, in a hunting and gathering mode of production. To use fishing as an example, the resource (salmon, herring, eulachon) concentrated in particular spots along the coast at various times in the year. The supply varied from year to year (sockeye salmon, for instance, run on four year cycles with two good years and two poor ones). Thus most units could realize some years of great abundance and some years of near starvation. The potlatch, universal among all tribes, served as a way to redistribute surplus among all groups. "In the last instance the potlatch was an institution for validating claims to resources, land titles, and the right to acquire surplus products from the use of clan lands" (Averkieva 1971, 334).

The institution of the potlatch enabled a wide variety of small economic units to become interdependent, exploiting the available resources. Although the technology was that of a hunting and gathering economy, the entire socio-economic system generated considerable surplus used to develop rich and diverse cultures. Economic inequality and exploitation existed and were more pronounced among certain tribes, but the entire economic organization depended on cooperation. Any external influence upset the equilibrium, and at best required adjustment within the system, as appears to have occurred in the first phase of European contact (Piddocke 1969).

The fur trade introduced an ideology of individual gain and prestige, as well as the means whereby an individual, rather than a group, could acquire wealth. Averkieva (1971) notes that inegalitarian relations were developing among the Tlingit. However, the Europeans influenced the direction of change. Moreover, European economic relations directly contradicted those established and practised by the First Nations. This was not evident during the initial contact period because the Industrial Revolution was just beginning to change the mode of production in Western Europe. The Europeans initially did not intend to develop resources other than fur and thus their population remained very small, confined to trading posts.

The trading company represented European power during the period of the fur trade. After amalgamating with the North West Company in 1821, the Hudson's Bay Company held a monopoly in what became known as British Columbia (Fisher 1977). The colony of Vancouver Island was created in 1849, when the British government granted title to the company. Eighteen fifty-eight, the year of the start of the gold rush to British Columbia and the Yukon, marked the transition from the predominance of the fur trade to a growing emphasis on settlement. That was also the year a second colony was created, mainland British Columbia. James Douglas resigned his post with the Hudson's Bay Company to become governor of both colonies. They were united in 1866, and in 1871 entered Confederation.

The second half of the nineteenth century marked an economic transition. The fur traders organized economic relationships oriented to overseas markets, and were interested in the colonial economy only in terms of its extractive capacities. Aboriginal peoples occupied an important position as resource suppliers. With the beginning of the gold rush, a new type of foreigner entered the colonies. Gold seekers had use for Indians only in subsidiary roles, often as guides or prostitutes. And for the first time the two groups competed directly over resources. Thus gold miners working the Fraser Canyon disrupted the salmon fishing and processing operations of groups on the river. The conflict led to violence. "Some of the miners came up from California boasting that they would 'clean out all the Indians in the land,' and there were instances of the kind of indiscriminate killing of Indians that was a feature of the American west" (Fisher 1977, 98). Unlike the American west, however, the Indians held their ground, and Douglas intervened to diffuse the situation.

The gold rush was short lived, but it represented the first stage of a new economic relationship between Europeans and the resources in the two colonies, one of direct access. Gold miners were a largely transient population, unlike the settlers who followed them. The settlers were interested in one thing, land. Access to it was mediated by European governors of the colony, who disposed of it under the assumption that it belonged to the Crown. "Absolute title (a European concept) has been vested in the Crown ever since Britain, Spain, Russia, and the United States, without consulting any Indians, settled the questions of sovereignty over this continent" (Duff 1977, 66, note). The British state recognized an obligation to the First Nations, but it was founded on the assumption that they were subjects. As aboriginal peoples themselves were to point out time and again, however, they never entered, either voluntarily or through armed struggle, into any such relationship with a foreign power. The concept of European title to the lands of British Columbia was imported by the Europeans who settled there. In other words, they simply assumed that the land was there for the taking. While the settlers and the provincial government denied any obligation to the original inhabitants, the colonial government under Douglas (but not his successors) and the federal government did recognize an obligation to "extinguish" aboriginal title by treaties and allocation of reserve lands. "The principles of the Royal Proclamation were followed by governments in Canada east of the Rockies. Aside from a brief beginning, they were ignored in British Columbia, and ultimately the province came to assert that the proclamation did not apply to its territory and had not been intended to. British Columbia Indians, however, came to regard the proclamation as guaranteeing their basic aboriginal rights. In their political thinking it gained the stature of a collective Magna Carta" (Tennant 1990, 11).

European concepts of private property relations to land and resources, especially those connected with capitalism, were foreign to the original inhabitants. Their concepts of ownership were based on acknowledged rights to use the land and its resources, rights that had to be negotiated by one group with its neighbours. And the rights carried with them an obligation to share surplus in return for recognition of title (for example, through the potlatch). However, the European code of rights was ultimately backed by armed force, and there are numerous instances in the various government reports of the use of armed vessels to "settle" disputes over land rights. The First Nations were at a disadvantage because their weaponry was inferior to that of the Europeans and there was no cohesive political institution that could negotiate collectively on their behalf in Victoria, Ottawa, and London.

One example of the use of armed force is provided by John A. MacDonald, who as Superintendent General of Indian Affairs wrote the following in his annual report dated 1 January 1886:

> In British Columbia the survey of the reserves on the North-West coast which were allotted to the Tsimpscheam Nation, in 1881, occasioned dissatisfaction amongst some of those Indians, they having become imbued with an idea, fostered among them by evil advisers, that they were the legal owners of the entire country, and that by permitting the boundaries of reserves for occupation by them to be surveyed they would weaken their claim to the title of the larger area. The surveyor who had been sent to the coast for the above purpose made repeated but fruitless efforts to accomplish the work; – the Indians on each occasion of his attempting to proceed with it, forcibly, but *without violence* – excepting on one occasion, when the surveyor's instrument was taken from him and his wrist was slightly injured in the struggle – prevented him from doing so. Representations having been made to the Provincial Government, the services of H.M.S. *Cormorant* were secured to convey a stipendiary magistrate, the chief of police at Victoria and a posse of constables to the locality. Upon the arrival of the vessel, arrests of eight of the ringleaders were effected. (DIA 1886, x–xi; emphasis added)

The fisheries were to play a crucial role in the conflict, since they were vital to the coastal aboriginal economies. For example, salmon canners made the first major capitalist endeavour in the north. The missionary William Duncan acted as intermediary on behalf of the northern Tsimshian. At his request, Indian Reserve Commissioner Peter O'Reilly travelled to the mission village of Metlakatla in 1881 to conduct hearings. The reason for the urgency was "the Indian fisheries were being taken possession of by whites for cannery purposes, and that if steps were not taken to secure to the Indians their fisheries, they would suffer great injustice" (British Columbia 1884, lxxvi). The Metlakatla Indians demanded the entire peninsula in question, including Turner's fishery on the Skeena, the location of Inverness cannery (the first constructed in the north) (Blyth 1991, 115–19). O'Reilly replied that he had no power to deal with lands sold to the whites. While he did his best to reserve the sites demanded by the First Nations, provincial administrators later claimed he had been too generous and cut back the size of many of the original allocations. In the end, settlers and capitalists received the major concessions. Salmon canners established their plants pretty well where they wished, occasionally running into disputes with local groups. Aboriginal peoples were assigned specific tracts of land and specific fishery sites. Thus the overall socio-economic system enforced through the potlatch was truncated. However, aboriginal economies were not destroyed, but continued alongside capitalist relations of production. As the original aboriginal system was dismembered, individual economic units became dependent on capitalist relations of production.

Equally instructive is the Report of the Government of British Columbia on the Subject of Indian Reserves submitted 18 August 1875, laying out the different policies of the Dominion government and the Crown Colony:

> The policy of the Dominion aims at a *concentration of the Indians upon Reserves*, while that of the Crown Colony, besides granting Reserves in cases where the Indians preferred them, courted rather the opposite result. The Colonial Policy was first inaugurated under the auspices of the Imperial Government in 1858, the date of the foundation of the Crown Colony. Under this policy the Natives were invited and encouraged to mingle with and live amongst the white population with a view of weaning them by degrees from savage life, and of gradually leading them by example and precept to adopt habits of peace, honesty, and industry. (British Columbia 1875, 58; emphasis in text)

While the federal government was pursuing a policy of segregation, the Crown Colony focused on assimilation, which many interpreted as neglect against the inroads made by aggressive white settlers who were taking their lands, as the report itself noted. Wage labour, which brought aboriginal people into contact with the "civilized races," was a chief means of assimilation. The language throughout this report makes it clear that there is a clear bifurcation between Indians or Natives, who are to be weaned "by degrees from savage life" (British Columbia 1875, 58), and the white race. The Natives are further divided into three "classes": fishermen and hunters, stock breeders and farmers on a small scale, and labourers. It was estimated that the first class comprised some thirty thousand "Coast Indians" (from an estimated population of forty thousand):

> No good reason exists why "Fisheries," such as those established by our merchants on Fraser River for curing and exporting salmon, and other merchantable fish, should not be erected in suitable places for the benefit of the Indians, and be in time profitably controlled and conducted by themselves.

Many of the Indians are now employed in this industry as fishermen, at one dollar, or four shillings sterling, a day. The business requires but little mechanical skill, and that they already possess ... The experiment might be made at a very small outlay, especially as all the necessary appliances – a few tools and some tinware excepted – are almost within their reach. In the comparative cost of labour they would possess an enormous advantage as long as wages remain at their present high figures. The Merchant, instead of embarking in such ventures himself, would doubtless find it more profitable to purchase his supplies from the Indian "Fisheries," which would thus at the outset be relieved of the responsibility of finding a foreign market for their goods. (British Columbia 1875, 63–4)

What George Walkem, the attorney general and author of the report, is suggesting here is that a policy similar to that of the fur trade be extended to the fisheries. This in effect would have embarked aboriginal peoples on the road to industrialization and given them control as capitalist entrepreneurs. While the missionary Duncan was to put this plan into effect at Metlakatla, the industry was founded and controlled by "whites" (see Tennant 1990, xi, for a discussion of the use of the term "Whites"). Aboriginal peoples became increasingly and chiefly seen as a source of supply of the raw fish or as cheap wage labour in the canneries.

What also emerges from this report is a view that "race" is not immutably fixed. In line with Miles' analysis, it would appear here that "the Other" can join "Self" if he gives up his savagery and acquires the habits associated with a civilized people. Since this was also the period of industrialization, "industriousness" is a virtue and habit associated with civilization. Thus class relations also emerge as determining factors in how "Self" and "Other" are characterized. Unlike gender, which is viewed as fixed by nature, the concept of race developed in this report is associated with "the white man's burden." It appears to be more fluid and permeable across developing class relations. And it is in all cases marked by an attitude already well developed regarding women: paternalism. Thus the missionary William Duncan, in a memorandum appended to the same report, offered this summary: "The Indians of British Columbia are by no means poor in the usual meaning of the word, i.e. they are not poor as to resources, but are ignorant, indolent, and improvident, and hence need a guiding hand before they can become a prosperous people" (British Columbia 1875). He recommends that all permanent Indian towns or villages be built as far away from the settlements of the whites as possible, "thus rendering them accessible to the Christian Missionary and Schoolmaster: for unless they become more collected it would seem impossible that education or civilization should ever reach them as a whole" (British Columbia 1875, 71). For Duncan, isolated reserves were ideal holding and collecting stations, bringing aboriginal peoples together to facilitate missionaries' indoctrinating them into a European, and Christian, way of life. This was clearly a one-way street.

Duncan acted on his beliefs and established the community of Metlakatla in 1862, removing aboriginal people from Fort Simpson, only to be expelled from the Church Missionary Society and to have a commission established in 1884 to investigate the land disputes that had led to the "severance." The various tribes living along the Nass and Skeena Rivers in northern British Columbia, and extending to the Queen Charlotte Islands, used the example of the missionary village to press for their own land claims and to expel white intruders.

To summarize the discussion thus far, examination of actual government reports and the analyses of state officials reveals a conceptualization of the "Indian" as "Other" that depends not only on the perception by the author of Indians as a group but also on that perception's being coloured by an evaluation of the economic needs of the province and the perceived role for Indians in its development – a development to be controlled by "whites." The conflicting definitions of property in terms of both land and resources and the rights that attach to its use are also notable. It was in the fisheries that whites directly intruded upon and contravened the prevailing property rights of the First Nations. While salmon canners left the definition of aboriginal rights to politicians, in practice they assumed their own direct right of access to the fisheries, while at the same time relying on aboriginal labour power and knowledge. Thus unlike mining, where Indians were ancillary to the activity or in direct conflict with it, in the nineteenth century fisheries they were indispensable. The role of the federal state in assuming control of the coastal fisheries becomes important here because it introduces yet another set of contradictory relations around property rights. Are the fisheries, according to European conceptions, common or state property (Marchak 1987, 3–31; 1989, 3–23)?

STATE CONTROL OF FISHERIES

Because fishing was so central to the way of life of aboriginal coastal peoples, it is appropriate at this point to examine in some detail how the fisheries were incorporated into the new nation state of Canada. The discussion also sheds some light upon the concept of property rights and the role of the federal state in defining what these mean.

The Department of Marine and Fisheries (see Canada. 1868–1925. Department of Marine and Fisheries, hereafter referred to as DMF) was created on 1 July 1867, although it did not acquire legal existence until the Fisheries Act was enacted the following year. Thus, upon the confederation of New Brunswick and Nova Scotia with Canada (the provinces of Ontario and Quebec), administration of the fisheries of Atlantic Canada was immediately assumed as part of the mandate of the new Canadian federal government. One of the chief reasons for such haste was the ongoing dispute over the right of fishers from the United States to fish in "colonial" waters. While the Reciprocity Treaty of 1854 extended to the fisheries on the Atlantic seaboard, many Canadians (among them, P. Mitchell, the first minister of the new department) considered that the Americans had abrogated that treaty. In a report dated 27 February 1868, P. Mitchell indicated that instructions had been received "from the Secretary of State for the colonies, dated 12th April, 1866, that Her Majesty's Government are clearly of opinion that by the Convention of 1818, the United States have renounced the right of fishing, not only within three miles of the colonial shores, but within three miles of a line drawn across the mouth of any British Bay or Creek" (*SP*, no. 12, 1869, p. 5). The Americans not only questioned the geographic application of the three-mile limit but also claimed their right to fish in Canadian waters as part of the reciprocity treaty.

Mitchell did not, however, view the matter as simply a question of legalities and/or application of the letter of the law. In his annual report for the year ending 30 June 1869 (DMF 1869, 51), he noted that the Americans "assumed weakness of Canadian jurisdiction. It is, therefore, fast degenerating into a virtual abandonment of all national rights of fishing on our own coasts. The time has arrived when we must either abandon this authoritative right, or assert and maintain it on the basis of treaties existing between Great Britain and the United States." On the following page he emphasized: "The temporary and indulgent policy hitherto pursued, will henceforth give place to a *definite policy of exclusion, agreeable to colonial interests and consistent with national dignity and rights*" (emphasis added). Clearly, the minister was staking the claim of the new nation of Canada to property that extended into the ocean and that incorporated the right to fish as a right of British subjects. He legitimated the right of Canada in this matter by reference to existing international law. C.B. Macpherson (1978, 3) notes: "As soon as any society, by custom or convention or law, makes a distinction between property and mere physical possession it has in effect defined property as a right ... What distinguishes property from mere momentary possession is that property is a claim that will be enforced by society or the state, by custom or convention or law."

Thus one of the first actions of the new minister was to declare the fisheries to be a form of property in terms of both which nations had right of access and who was to be excluded from use. This was so fundamental and important that there was talk of warfare. By 1869 a force of Marine Police had been established to "guard British fishermen against molestation" (DMF 1869, 50). Although Mitchell viewed the potential of the Atlantic fisheries as "exhaustless," he was determined to establish boundaries and enforce them with armed might if necessary. Mitchell noted in his report: "Immense as is the intrinsic value of the exhaustless fisheries which form so large a portion of our natural resources, their rightful control and exclusive use possesses a peculiar value and significance intimately connected with the new conditions and prospects of this country" (52).

A letter from W.G. Romaine, Admiralty, to the under secretary of state dated 4 June 1869 dramatized the seriousness of the situation. In it he stated: "It is certain that the United States will send vessels of war to look after their fleet of 700 fishing vessels in the waters of the Dominion" (*SP*, no. 81, 1870, 6). In response to this perceived threat, the marine police force was expanded. The following season marine police boarded 400 American fishing vessels. Mitchell was quite blunt in his recognition that, in the end, it was not so much a question of the ability of American fishers to use what were established as Canadian waters as it was a matter of asserting the new nation's right to territory over and against the established power to the south. "The fact is, that American fishermen cannot successfully prosecute the Gulf Fisheries without using our inshore waters, nor without frequenting our bays, ports and harbours ... The popular idea that we do not admit them to a free use of our inshore fishing grounds, either in retaliation for abrogating reciprocity, or in order to coerce the United States Government in the direction of reciprocal free trade, is an erroneous assumption. We simply deny them the continuance of valuable benefits for which they have ceased to afford us any equivalent" (DMF 1870, 65–6).

Further on in his report, Mitchell refers to "the teeming waters around the coasts of the British North American possessions" as a "national property richer and more perpetual than any mere monied estimation could express" (DMF 1870, 69–70). Later in the same report, he asserts: "We claim for our people the same exclusive right to control the fisheries within the three mile limit that we do to control the mines, or the timber, or the soil of our country." In the end, the entire matter was put to a special commission of inquiry, and in 1871 the Treaty of Washington was ratified in London. The U.S. Congress was tardy, however, and it was only on 1 July 1873 that the Fishery

Articles of the treaty came formally into operation. Article XVIII gave Americans the right to take fish with no restrictions on distance "provided that, in so doing, they do not interfere with the rights of private property, or with British fishermen" (*SP*, no. 18, 1872, 36). It was further recognized, in Article XXI, that as the United States benefitted more there should be compensation. Again, the United States dragged its heels in coming to an agreement on this matter.

While British Columbia did not enter Confederation until 1871, the fact that a department had already been created and that political and territorial muscles had been flexed was to affect in a very direct manner the exploitation of the fisheries on the Pacific coast. Upon entry, the B.C. fisheries were added to the mandate of the Department of Marine and Fisheries. Item 11 of the "Proposed Terms of Confederation with the Dominion of Canada" read: "Whatever encouragement, advantages, and protection are afforded by the Dominion Government to the Fisheries of any of its Provinces, shall be extended in similar proportion to British Columbia, according to its requirements for the time being" (*SP*, no. 18, 1871, 7).

James Cooper, marine and fisheries agent for British Columbia, noted in his report dated 30 January 1874 that the "benefits arising from the Washington Treaty are anxiously hoped for in British Columbia" (*SP*, no. 4, 1874, 206). In particular, it was anticipated that fish products could enter American markets on the Pacific coast duty free. However, U.S. customs officers at San Francisco and other ports denied "the right of the Province of British Columbia to participate in the terms of the Washington Treaty which provide for the admission free of duty of fish and fish oil, &c." (*SP*, no. 42, 1876, 3, letter dated 31 March 1875 from Isaac Burpee, Minister of Customs, to the Governor in Council). The treaty was signed 8 May 1871, while B.C. entered Confederation 20 July 1871; however, although the act of parliament giving effect to the Treaty of Washington in Canada only passed on 14 June 1872, the request was in the end denied.

With the election of a new government, in 1873 the Honourable Albert J. Smith took over as Minister of the Department of Marine and Fisheries. W.F. Whitcher was appointed Commissioner of Fisheries, and in his report for the year ending 31 December 1875, he included a section entitled "Riparian Claims," which further clarified the relationship between the state and property as concerned the fisheries. He noted that the courts of Ontario, Quebec, Nova Scotia and New Brunswick "have invariably maintained that the soil of public navigable waters belongs to the Crown, and the right of fishing belongs to the public. Private individuals have, on numerous occasions, claimed them both, but have entirely failed to establish such claims" (DMF

1875, xxxvi). On the following page he remarks in relation to the 1868 Fisheries Act: "The whole tenor of that statute is an authoritative denial of any other private claims to fishing privileges, either absolute or incidental, express or implied, in the public navigable waters of the Dominion":

It has always been accepted and acted upon as a recognized principle in administering the Fishery Laws, that the fisheries are a public property which the Crown is now empowered by Act of Parliament to control temporarily, but not in any case to alienate. In exercising this authority the leading object of all concerned has been to preserve and improve these public fisheries. The next aim has been to promote the interests of practical fishermen, and to protect them in the just use of the fishing privileges secured to them by Common Law. (xxxvii)

Fisheries experts have long debated the relationship between the state and the fisheries in terms of property rights. Are the fisheries common property, state property, or private property managed by the state? Regardless of the particular position one chooses, it is clear that from Confederation the federal state began to regulate not only the management of the fisheries as common property but also access to them, as the clash between the Crown and the U.S. government demonstrates. With the incorporation of British Columbia into Confederation in 1871, the mandate of the Department of Marine and Fisheries was extended to the Pacific coast. However, it was recognized from the first that the laws and regulations developed for the Atlantic did not necessarily apply to the Pacific fisheries. And from the beginning, the different population base in British Columbia was taken as a chief factor in regulating fishing and promoting its commercialization and industrialization.

Governor Musgrave wrote to the Governor-General on 20 February 1870 on the subject of the proposed union. "The true number of the population is not known, and it includes a large number of Indians, who are to a great extent consumers. The white inhabitants are chiefly male adults of wasteful and expensive habits. The production of the Colony is very small, except of gold" (*SP*, no. 18, 1871, 2).

In his address to the Legislative Council in Victoria on 15 February 1870, Musgrave noted that $5000 had been appropriated to promote immigration of women and of agricultural settlers. He was looking to England as the chief source of migration. Thus even before its entry into Confederation, the politicians who would govern it had typed the population of British Columbia according to race and gender.

Before examining the nature of these classifications in more detail, it is worthwhile to keep in mind the warning against conceptual inflation that Miles develops in his book, *Racism* (1989). While the concept of race is a necessary precondition to the development of an ideology of racism, the two are not equivalent. In the early years of the establishment of B.C. as a province, this becomes quite evident. While some politicians and employers, as well as members of the public, clearly expressed racist views in their denigration of nonwhite populations, others developed a more tolerant attitude, although it was generally tinged by a paternalistic evaluation, especially of aboriginal peoples. A good example of this appears in the annual report of the minister of public works, dated March 1872. The Hon. H.L. Langevin visited British Columbia in order to assess its requirements, especially in regards to railway construction. But his report is far broader in scope. Thus, in a section entitled "Fisheries" he notes: "The fisheries of Columbia are probably the richest in the world, but they have been but very little worked. The gold fever draws immigrants towards the auriferous tracts .. the fisheries require fresh arrivals to develope [*sic*] their full resources ... there are really only two large fishing establishments: one a salmon fishery under the management of Captain Stamp, who, for the first time, exports salmon in tin boxes; the other, a whale fishery in the Gulf of Georgia" (*SP*, no. 10, 1872, 15).

On the following page he quotes from a memorandum he received "from a gentleman who is ... in a position to give the correct information on this subject" that in "speaking of the fisheries of British Columbia, one may almost be said to be speaking of something which has no existence. With the exception of a small attempt at putting up salmon in tins on the Fraser River, and one or two whaling enterprizes of a few years standing, no attempt whatever has been made to develope [*sic*] the actually marvellous resources of this Province in the way of fish." Missing from this account is any acknowledgment that fishing was a central economic activity pursued by aboriginal peoples. Clearly, what Langevin had in mind was the commercial and industrial prosecution of the fisheries, and this formed the basis for his reporting.

Langevin noted that a census had been conducted in March or April of 1871: 8576 whites; 462 negroes and 1548 Chinese for a total of 10,586. The Indian population was excluded from this census. In addition, Langevin took note of the "usual disproportion between the two sexes, which was remarked from the first in California and Australia. Thus the number of men is 7,574 and of women, 3,012" (*SP*, no. 10, 1872, 22). Langevin categorizes the population not only by sex but also by race. The use of the term "white" appeared to be widespread. For example, he also remarks that the "white population is composed in great part of men of education many of whom have seen better days" (22). A section of his report is entitled "Chinese":

> The Chinese population is regarded with no greater affection in Columbia than in California, but is, at any rate, in the former country, not ill treated. The Chinese are an industrious, clean and laborious community. They are generally miners, working either on their own account or for the whites, or in service as laborers or domestic servants, most frequently in the latter capacity as cooks. Many of them are also employed, either on their own account or by others, in the transportation of freight from the lower Fraser to the mines. They live on little, and make a livelihood even in places from which the whites have had to emigrate lest they should perish of hunger. Those who dread the increasing immigration of the Chinese maintain that they are a population but little to be desired. They forget that the presence of the Chinese has contributed to reduce the price of wages in the mines. A Chinese there receives \$3 to \$3 50 a day, and a white man \$5. (23)

This paragraph is worth quoting in full because it demonstrates not only classification by race but also class relations, particularly the employment of Chinese as cheap wage labour. The racism expressed by white miners and others was influenced by their resentment of Chinese labourers undercutting and undermining their own wages. It was in the interests of employers to fan this division since it served to deflect resentment onto labourers who were disadvantaged and to keep the various segments of the working class from uniting and identifying employers as responsible for differential wages. But employers in turn were acting within a pre-established ideology that allowed them to use the categories of sex and race to determine which group could be hired for cheap wages in the first place.

Langevin appends a memorandum written by Chief Justice Begbie describing the aboriginal population. Begbie notes that the "coast Indians" had been at least three times as numerous fifty or sixty years previously. He attributes the decrease primarily to diseases such as small pox and measles and, to a lesser extent, to warfare and "whisky and dissipation" which "wear them to an early death, and sap the powers of reproduction." He notes that vaccination against measles is "as severe a malady as measles among Europeans." Elsewhere in his report, Begbie refers to the Indians as "savages," although he also admires their skill in the manufacture of canoes. His next comment is especially telling, since he distinguishes between intelligence and the ability to reason in an abstract manner, which he links to religion. "Alone among all intelligent savages of whom I have read, (for there is no doubt these are very intelligent, and with vast natural power of observation,) they seem to have no religion whatever … I have never heard of an Indian language which possessed in its vocabulary a word expressive of an abstract idea" (24–8). Begbie also submitted to Langevin a "Dictionary of Chinook Jargon" which Langevin attached to his own report as Appendix CC (*SP*, no. 10, 1872, 161–82). In her history of British Columbia, Barman (1991, 169) provides the following information about the importance of Chinook by the second half of the nineteenth century. "The persistence of everyday relationships between Europeans and native peoples is embodied in Chinook. Emerging out of early contact, the Chinook jargon possessed at most seven hundred words, derived in approximately equal proportions from the powerful Chinook Indians of the lower Columbia River, from the Nootka of Vancouver Island, and from French and English ... With a knowledge of Chinook Europeans were able to communicate with native peoples speaking very different languages ... Chinook was widespread across British Columbia, for a time being possibly spoken and understood by more people than any other single tongue, including English."

Begbie states that while he is not aware of any treaties having been made, reserves have been laid out on both Vancouver Island and the mainland (26). He does not think any general treaty would be possible because of the vast number of tribes who cannot communicate with one another due to language barriers and mutual hostility. He advocates dealing fairly with them in the matter of settling land claims. However, his approach is thoroughly paternalistic, as the following statement makes clear. "They are in that state of powerlessness and respect for the superior power, numbers, and acquirements of the governing race, that any arrangements which that race would, consistently with self-respect and humanity, think proper, would readily be adopted by the native" (26). Begbie notes that their system of land tenure, by which village communities occupy and cultivate irregularly detached plots, is "scarcely intelligible to English notions of property in land." He also remarks that "they have an affection for particular little bits of land, (which seems a feeling common to humanity, savage or civilized), which, probably, is exceedingly inconvenient to a surveyor, and is not always, in our view, very reasonable. It is, in fact, prejudice." He concludes that the Indians are "very valuable inhabitants of the Province" and should be taught settled habits, and, in particular, agriculture along the lines followed by Mr Duncan at Metlahkatlah (27).

Begbie goes on to elaborate why he calls them valuable inhabitants. First, they helped to open up the country for gold miners. "Until roads were made, no supplies were taken in except by Indians." Second, they are "large consumers, in proportion to their means, of customable articles." Third, they are "our very best tools" for obtaining furs. "But these qualities are exactly those which make it very difficult to civilize them ... Accordingly, after years of cultivation, he constantly relapses, for a time at least, into a painted savage, and goes hunting and fishing – or starving – as a relaxation" (27). Missing entirely from the discussion is any acknowledgment of the pre-existing cultural context within which such activities would make sense. Begbie's viewpoint is entirely ethnocentric. It was this type of mental construction that could be used to justify paying them starvation wages. One could reason that, after all, they were used to living on little and probably would not know what to do with more money except to "dissipate" it in whisky and debauchery.

Langevin picks up from Begbie's report and confirms the view that the Indians are an important population "in the capacity of guides, porters, and laborers." Indians have learned to regard the person of the British subject, "King George Man," as "sacred," although this respect does not extend to Americans (28). In the next section of his report, "Their Progress Towards Civilization," Langevin notes that some of the children, including girls, are receiving education at mission schools (29). Very few of them, however, are Christians, which seems to be a primary criterion for the road to civilization. It is important to note, however, that both men seem to see the possibility of "civilizing" the savage, although whether this would ever qualify the Indian as a person of the same order as a white man is a debatable point. Both authors relegate Indian men (they rarely mention women) to menial jobs for which they are seen as particularly suited by nature. Neither man entertains the idea that Indians could become politicians or capitalists. In Carstens' terms, "Indians" are clearly set apart within "little colonies." Indeed, the main concern of both Langevin and Begbie is that aboriginals stay put on the pieces of land reserved for them and that the white man exercises "humanity" and "justice" in allocating the land to them. It is never suggested that land claims involve negotiations between equals. Thus, the transformation of the idea of race toward an ideology of racism is present, even if it is not expressed in exactly those terms.

Equally important is the development of an independent nation state. The "founding fathers" of Confederation had a vision of transforming the country into a nation on the European model. This vision in turn was already formulated by and steeped in relations of hierarchically organized power and excluding groups by their race and gender. At the same time, an alternative vision of state was developing in British Columbia, a vision steeped in the same traditions but with a different set of power relations *vis-à-vis* the federal state. Barman (1991, 346) draws attention to the argument developed by historian Keith Ralston that in the first decade and a half of Confederation community leaders developed ties that linked them more closely to the rest of the world by means of overseas routes from the Pacific than to the rest of Canada. The industrialization of the fishing industry began in this field of relations. Markets were established via Victoria overseas to Great Britain and the other dominions. And labour forces were drawn from international migrations of Chinese peasants and Japanese fishers whose migratory routes initially took them to the Pacific coasts of California, Oregon, Washington, British Columbia, and eventually Alaska rather than horizontally inland. The place of Chinese shoreworkers and Japanese fishers as cheap wage labour is explored in the next chapter.

ORIGINS OF SALMON CANNING IN BRITISH COLUMBIA

Colonization and pre-existing patriarchal relations in Western Europe influenced the categorization of groups of people according to socially constructed criteria of race and gender. Underlying this system of bifurcated thinking was a material reality that structured labour capacity in unequal relations of oppression while enabling the physical maintenance of the subordinated groups at a level below that of the groups in power. Capitalist class relations and the process of industrialization then seized upon and transformed these relations in new ways. While one dimension of this process was colonization of indigenous and overseas populations and their transformation into seasonal wage labourers, another equally important facet was the technological implications of industrialization as these came to be applied to primary resources like the capture, processing and marketing (in other words, the commodification) of fish. (See Tables 1–3.)

All the indigenous coastal tribes had a variety of processing techniques and a division of labour that assigned this work to women. Fish being "the basis of Northwest economy" (Drucker 1958, 35), it was essential to preserve it for winter use. With industrialization, however, while the division of labour by which women did processing work was retained, the indigenous methods of preservation were not adopted.

When the fur-trading companies became land based, they salted fish for export to crews stationed elsewhere. According to Rounsefell and Kelez (1938, 701), around 1800 the Northwest Company began to salt salmon, and the technique was then adopted by the Hudson's Bay Company. By 1835 HBC was shipping three to four thousand barrels of salted salmon yearly to the Hawaiian (Sandwich) Islands. Dried salmon was also important in provisioning those stationed at the forts.

Table 1
Origins of the people, British Columbia, 1881–1901

	1881	*1901*
BRITISH*		
English	7,297	52,863
Irish	3,172	20,658
Scotch	3,892	31,068
Welsh	299	not specified
Others	not specified	1,814
African	274	532
Chinese	4,350	19,482 (Chinese and Japanese)
Dutch	94	437
French	916	4,600
German	858	5,807
Indian	25,661	25,488
Italian	143	1,976
Jewish	11	534
Russian and Polish	48	1,143 (Russian only)
Scandinavian	236	4,880
Spanish and Portuguese	144	not specified
Swiss	40	249
Austro-Hungarian	not specified	1,377
Belgian	not specified	410
Half-breeds	not specified	3,461
Various other origins	342	479
Not given	1,682	1,390

Sources: The Census of 1881, Table III-Origins of the People, 298–9. *Report on the Fourth Census of Canada, 1901*, Table XI. Origins of the People, 284–5.

* The 1901 census grouped those of British origin together and placed them first, unlike the census of 1881 which followed alphabetical order and separated those of British origin according to country of origin (English, Irish, Scotch and Welsh).

"At the Post of Fraser's Lake, in 1836, 36,000 dried salmon were purchased and stored for use; and at other posts proportionate quantities were likewise secured out of the superabundant provision made by the natives ... At Fort Langley (some 15 miles above New Westminster) large quantities were formerly salted every year by the Hudson's Bay Company, as well for home consumption as for exportation. In some seasons between two and three thousand barrels were thus provided; the fish procured by barter from the natives" (DMF 1873, 184).

But while the trading companies depended upon salmon for their food supply, there was no larger market for this type of processed fish to allow its commercial development.

Table 2
Sex and Conjugal Condition in British Columbia, 1891–1901

	Year	*Male*	*Female*
Sex	1891	63,003	35,170
	1901	114,160	64,497
Single	1891	42,580	18,471
	1901	75,093	35,274
Married	1891	18,811	14,809
	1901	36,429	26,090
Widowed	1891	1,612	1,890
	1901	2,586	3,096
Divorced*	1901	52	37

Source: Census of Canada 1890–91, Table III, compared by Census Districts, 10–11.
Note: Table III-Civil Condition, pages 116–19, gives the number of families in British Columbia in 1891 as 20,718 out of a population of 98,173. The average size of a family is 4.7. This compares to 5.2 for Canada as a whole, which had a population of 4,833,239 and 921,643 families.
* This category appears only for 1901.

Just as lumbering created its own social structures, so the young industry of salmon canning brought a distinctive set of relationships into being. Salmon had always been valued as a food in the Pacific Northwest. Fur traders soon replaced the traditional pemmican with salmon as their diet staple, each man allowed so many dried fish a day as his food ration. The HBC pioneered the export of salted salmon in barrels to markets in Hawaii and as far away as London. Britain in particular was eager for any available food products. As a prime industrial country it simply could not produce the food necessary to feed its population. Yet salmon possessed limited commercial potential as long as preservation techniques were uncertain. (Barman 1991, 117)

The introduction of a fully capitalist enterprise depended on developments in processing techniques. The technique of canning food enabled the industrialization of the food industry, and in turn facilitated the growth of an urban population lacking the means to feed itself. Developments in transportation were also indispensable in facilitating movement of raw materials to processing sites and of manufactured goods to the urban masses. "The canning industry had come into being in the 1840s with the development of stamping and forming machinery for producing tin cans on a mass basis" (Braverman 1974, 262). However, the canned food industry did not expand to incorporate national and international markets until the 1870s, requiring further technical developments as well as new developments in rail and sea transport.

Table 3
Manufactures Fish, preserved, in the Province of British Columbia, 1901

No. of Establishments: 70

Values of real estate, plant and working capital

Lands	*Buildings*	*Machinery and motive power*
$217,133	$910,226	$300,279

Salaries and wages of persons employed

	Number	*Salaries/Wages*
Owners and firm members	91	$87,926
Salaried officials, salesmen, etc.		
Men	117	$107,688
Women	3	$935
All others	2,481 (men)	$349,700

Aggregate working days in year

Men	*Women*	*Children (under 16 years)*
191,984	50,582	4,300

Source: Fourth Census of Canada 1901, Volume III, Manufactures. Figures taken from Table I (p. 2); Table II (p. 18); and Table III (p. 40). In 1901, the average wage earnings per day in British Columbia were $2.12 for men; $0.98 for women, and $0.67 for children (Table 6, p. xiv). While the wages for all three categories of labour were amongst the highest in Canada, the category of "Fish, preserved" had the second lowest ranking for Canada as a whole when it came to "Average earnings for labour per worker": $78.42 compared to $68.33 for "Evaporated fruits and vegetables." The highest in the "Food products" group were $600 ("Sausage casings"); $583.33 ("Slaughtering") and $510.57 ("Sugar refining"). The "Average earnings for labour per worker" in Canada in 1901 was $189.44 for "Food products" as a whole compared to $238.77 for "Textiles" (Table 33, p. xxxviii).

Thus from the start salmon canning in British Columbia was linked to a variety of national and international developments: in processing, in marketing, in transportation, and in financing. Having begun on the Atlantic coast (Maine and New Brunswick), salmon canning was introduced on the Pacific Northwest coast, and "salmon canneries spread in about twenty years from the southern limit of salmon habitat in the rivers that flow into San Francisco Bay to the northern limit in Alaska, leapfrogging in a frenzy of development from the Sacramento to the Columbia, from the Fraser to the Skeena, and finally into the rich salmon streams of Bristol Bay, Alaska" (Ralston 1981, 299).

It is not clear who canned the first Pacific salmon. Credit is generally given to George and William Hume, known to be canning in 1864 on the Sacramento River, "but British Columbians claim the Humes were preceded by at least four years by Captain Edward Stamp, the province's 'first industrialist'" (*Pacific Fisherman* 1952, 16). Cobb (1930, 429) also credits William Hume as having helped in the establishment of the first salmon cannery in the United States in 1864, but finding the run of fish on the Sacramento River disappointing, he started a cannery for Hapgood, Hume and Co. on the Columbia at Eagle Cliff in Washington. In 1881, of the thirty-five canneries operating on the Columbia River, Cobb estimates that about half had been established by the Hume brothers (of which there were four). Cobb (430) also notes: "George W. Hume was the first salmon canner to employ Chinese. This was at Eagle Cliff in 1872. At this period the white laborers in the canneries were recruited from the riffraff and criminal element of Portland. He had a Chinese working for him and through this man secured a Chinese gang from Portland. This labor proved so satisfactory that the custom soon spread to the other canneries. It was not found that the Chinese could do the work any better or quicker than the white laborer, but they proved more reliable in their work and gave less trouble."

Cobb credits Alexander Loggie & Co. for first undertaking the canning of salmon as a business on the Fraser River. In June 1870 Loggie built the first salmon cannery in British Columbia (according to Cobb, that is), located at Annieville. This was, apparently,

> A very primitive affair; the cylinders upon which the cans were shaped were of wood covered with sheet iron; the trays were small wooden contrivances holding about three dozen one pound cans. [There was] practically no machinery; the operations were almost entirely by hand. The fish after being put into the cans were preserved by boiling in large wooden vats. Great difficulty was experienced in thoroughly cooking the fish, the boiling point of ordinary water not proving sufficient ... salt was added to the water, and by this means the temperature was raised ... no windows or doors were allowed to be opened, except of necessity, under the mistaken idea that the cold currents of air would injure the product. (Cobb 1930, 471)

Captain Stamp also entered the business, and his cannery was located at Sapperton in New Westminster. He did not try to make his own cans but rather purchased them from Mr Deas, "a tinsmith of Victoria." Deas also entered the salmon canning industry, establishing a cannery on the island in the Fraser which bears his name, Deas Island. Ralston (1977, 72) claims that in these early years "Deas was the leading canner on the Fraser River." Deas (1838–1880) was "described as a mulatto ... In Victoria Deas joined a black community of several hundred, started in 1858 by organized migration from San Francisco" (216). The community was established when racist legislation was in process of being enacted in the United States but there was a general exodus back to the States after 1865 when the gold fields declined and conditions in the United States improved.

Ralston (1977) admits that it is difficult to know the extent to which Deas personally suffered from discrimination. At the height of the 1877 fishing season, Ewen and Wise laid a charge against him that he had used "violent and threatening language" to their fishermen (74). Deas was sentenced to three weeks in jail, but the case was later thrown out of court. However, the strategy worked in that Deas lost much precious time at the height of the season. Not long after this incident he left the canning industry after having engaged in it for seven seasons. Ralston also points to the fact that in reporting on the charge, while the New Westminster newspaper generally supported him and reported he was well liked, at the same time referred to him as only the manager, not the owner, of the cannery. "Other instances of people regarding Findlay, Durham and Brodie as the real owners, and not just Deas' agents, also occur. It was just too difficult to believe that a Black could be a major entrepreneur in an industry whose success was so devoutly wished for" (78).

Since the industry began in the United States, Americans came to dominate markets for the commodity (Ralston 1981, 300). They established a market for red kings, a species unknown in Canadian waters. B.C. canners were forced to find a species similar in colour, texture, and taste if they wanted to compete for the same market. Initially experimenting with red springs, they settled on sockeye salmon, the Fraser River being the largest sockeye-producing stream in the province.

Thus James Cooper, agent of the marine and fisheries department for British Columbia, noted in his report for 1873 that, while salting salmon in barrels had been extensively prosecuted on the Fraser River for fifty years, "originally by the Hudson Bay Company, and latterly by private individuals," canning salmon "is now developing every year." The firm of Messrs Findlay, Durham & Brodie canned 115 tons of salmon that year while other parties canned eighty tons. The canned salmon was exported to Great Britain while the salted fish was sent to the Sandwich Islands and Australia. Cooper advises against extending the fishery laws "in their entirety" to British Columbia since they "would probably lead to complications with the Aborigines" (DMF 1872, 206). The Fisheries Act was in fact extended to Prince Edward Island, British Columbia, and Manitoba the following year (1874), but the provisions of the act were not applicable in detail in those provinces.

In 1874 four establishments were operating on the Fraser with more firms preparing to enter the industry. Messrs Findlay, Durham & Brodie canned 7119 cases of fresh salmon and 254 barrels of salted salmon; Messrs Loggie & Co. put up 6500 cases of fresh salmon, 1000 barrels of salted salmon, and 100 barrels of bellies; Vancouver Island Co. reported 3000 cases of salmon and 120 barrels of salted salmon; Messrs Holbrook & Cunningham had 2100 cases of fresh salmon, 300 barrels of salted salmon, and 100 half-barrels; finally Mr Frederick Kaye salted 300 barrels of salted salmon, and "other parties" were estimated at 500 barrels. Each case of canned salmon contained 48 tins with one pound in each tin. While canning fresh salmon was gaining in importance, whaling in island waters had been completely abandoned because the company prosecuting this "fishery" lacked "proper appliances and the necessary capital" (DMF, 1875, 169–70).

In his report for 1876 Alex. C. Anderson, the new inspector of fisheries for British Columbia (appointed in May at a salary of $600 a year), reported only three firms prosecuting the salmon fisheries on the Fraser, but expected others to enter shortly. Compared to previous years, the price was very low. The reason given points to one of the endemic problems of the B.C. salmon canning industry: competition, in this case from the "enormous supply thrown into the market from the 'Canneries' on the Columbia River" (DMF, 1877, 340). A related problem was the inconsistent nature of the return of the beast. In good years there were huge runs with much competition, while in poor years the supply shrank, with the result that firmly established companies could buy up the competition in anticipation of better years and the desire to control at least the local production. Given the need to closely regulate the canned product because of botulism, the commodity tended to be uniform in quality. One strategy the canneries developed was to introduce colourful labels and brand names which they then flogged in the marketplace, by the mid-twentieth century vying with one another for a place on the grocery shelves of major retail chains both in Canada and abroad.

In contrast to most of the DMF officials who succeeded him, Meggs (1991) and Newell (1993), among others, view Alexander Caulfield Anderson (1814–84) as supportive of the rights of aboriginal peoples to the fisheries. In a short biographical sketch, Lamb (1982, 16) charts his career in North America as beginning in 1831 when he joined the Hudson's Bay Company. He had charge of several posts in the Pacific Northwest, and in 1858 he arrived in Victoria where he was appointed postmaster. "In 1876 he was appointed dominion inspector of fisheries with jurisdiction over British Columbia coastal and inland waters. The same year the federal government asked him to act as its member on a dominion-provincial joint commission on Indian land in British Columbia. This proved a frustrating assignment for Anderson because the efforts of the commission to delimit Indian reserves were defeated by the hostility of the provincial government" (17–18).

North of the Fraser, near the mouth of the Skeena River, another fishery was established that summer of 1876, known, Anderson thought, as the North-West Fishing Company, originally of San Francisco with capital also partly raised in Victoria. However, the selection of the station was a poor one and the company did not meet with the success that it had expected. The actual history of Inverness Cannery goes back a few years. The cannery site was part of Indian Reserve Number 6, and in 1870 William Woodcock, "who had no legal title to the property, had established a public inn and trading post on the site, which he called Woodcock's Landing. This was to accommodate miners and fur traders travelling the Skeena River" (Blyth 1991, 115). In 1871 a Crown grant was issued to Henry Soar and was transferred in 1874 from him to Drake, Findlay, and Wilson. The following year a Colonel Lane from Columbia River arrived on behalf of Victoria businessmen to scout the possibilities of establishing a salmon canning plant on the north coast. In 1876 the site was transferred to the North Western Commercial Company which "built the first salmon cannery north of the Fraser River and named it Inverness. They began with 40 fishboats and 225 employees" (115). Table 4 provides a list of the number of canneries by location for the years 1871 to 1902 as compiled by Cicely Lyons who worked for B.C. Packers and wrote a history of the industry.

The second cannery built on the north coast was Aberdeen (also known as Windsor since the Windsor Canning Company owned it). Salmon canneries were prone to destruction by fire, given their wooden construction, the need for intense heat to cook the contents of the cans, and the high oil content of the fish. Thus in 1893 Inverness was completely destroyed but immediately rebuilt. It canned salmon continuously for 107 years. In 1880 Turner, Beeton and Company took over its ownership, selling it to J.H. Todd Company in 1902, who retained ownership until the 1950s when the two big giants, Canadian Fishing Co. and B.C. Packers extended their grip on the industry.

In 1879 the two canneries produced 10,603 cases of 48 pounds each of canned salmon. The fish caught in the area was purchased from Indian and white fishers, the latter being supplied with nets and boats by the companies and paid 1 cent per pound. Inspector Anderson notes, however, that the Indians supplied fish at a cheaper rate. "At this point, as elsewhere throughout, the services of the Indians are utilized in the work of the fisheries. At Inverness, in addition to the

Table 4
Number of Canneries by Location, 1871–1902

Year	*Fraser River*	*Skeena River*	*Nass River*	*Central Coast*	*Vancouver Island*	*Total*
1871	1	–	–	–	–	1
1875	3	–	–	–	–	3
1880	7	2	–	–	–	9
1885	6	2	–	1	–	9
1890	16	7	3	6	–	32
1895	21	7	1	6	1	36
1900	42	10	1	11	1	65
1902	42	10	2	12	1	67

Source: Cicely Lyons, *Salmon: Our Heritage* (1969, 146–7, 164, 705–6). Canneries underwent a great number of changes in terms of ownership and location. Sometimes a cannery would be renamed when a new owner took control, while at other times the old names were retained. This makes it difficult to ascertain with any great precision the actual number of canneries in operation in any given year (canneries did not always operate every season). These figures are approximate and should be read as an indication of over-all trends, especially the escalation in the numbers built at the turn of the century on the Fraser, Skeena, and along the central coast. The year 1902 marked B.C. Packers' first bid to establish control of the industry by buying out owners and closing plants.

open-air laborers, a number a [*sic*] women were employed in the net-loft making salmon-nets, an art in which they are expert" (DMF 1880, 283–4). Anderson also noted that the Indians employed were citizens of "Metlah-Katlah" (285).

Blyth (1991, 92) notes that Metlakatla was founded in 1861 by William Duncan of the Church Missionary Society of England:

> His major goal was to instruct and train the aboriginal peoples of the north coast to become morally and economically competent to hold their own in the sudden encroachment of European development and culture.
>
> Duncan insisted that Christianity and the Church become the central influence in the lives of the Native people. From that base he mobilized and expanded economic development in such trades and industries as blacksmithing, weaving, making bricks, manufacturing products such as soap, tanning hides, and operating their own sawmill to provide lumber for themselves and for marketing. Part of Duncan's plan to develop a stable community was the building and operating of a salmon cannery.

Metlakatla Cannery was built in 1882, and it operated for three seasons. In 1887 it was dismantled. Duncan ran into trouble with his superiors in England and relocated almost the entire village of 1200 people, including the cannery, to Alaska. Blyth (1991, 93) notes that it was "the first all-Native Indian salmon canning plant to be constructed, owned, and operated on Reserve lands. Except for the management, the plant was run by the Indian people on a co-operative basis." However, it is interesting to look at the questions the commissioners posed to Duncan in 1884. It would appear that the cannery was Duncan's idea and that he regarded it as his property; it is certain that he saw himself as its manager. It appears that some of the residents of Metlakatla had not favoured building a cannery out of the proceeds of the profits made from the village store.

While 1876 was generally a poor year, the following year reversed the trend. Fish products in British Columbia were valued at $583,432.76 of which $436,667.76 was attributed to salmon in cans, showing the increasing predominance of salmon canning in the fisheries of the province (DMF 1877, xi). There was an extension in the amount of invested capital as well as a heavy run of salmon up the Fraser River. Allegations were also made that the Indians destroyed millions of salmon, but Anderson refuted these charges: "Their modes of catching the salmon in the upper waters – to them a necessity for food – is inobjectionable [*sic*]. Any interference with the natives, therefore, under hastily formed judgment or frivolous pretext, would be imprudent as well as unjust" (289). The product of the B.C. fisheries was valued at $925,766.98 for 1878, of which $736,138.80 was attributed to canned salmon. In 1879, the overall fisheries of B.C. produced $631,766.64 in revenue, including $395,882.64 from canned salmon. It is interesting to note that included in the annual report was an estimate of the home consumption of salmon by the Indians of B.C. for 1879. The population was estimated at 35,000 and their consumption of fish was estimated at $4,885,000, of which $4,375,00 was attributed to the consumption of fresh and dried salmon, almost ten times the revenue from canned salmon for that year (Supplement No. 2, DMF 1879, viii)! This estimate (which was not changed) appeared in the preface to the annual report over the next few years, indicating that the department regarded the food fisheries of the aboriginal peoples as related to the industrialization of the fisheries, particularly salmon canning.

The canneries as well as the aboriginal peoples were given bad press for wasting fish. In these early years, canneries dumped offal directly into the Fraser River, polluting it and creating quite a stink, literally as well as publicly. In addition to the fact that the companies threw away unwanted portions of the fish (the reduction industry was to partially solve this problem) this points to yet another problem, especially prevalent in the early years: hot weather combined with large runs often meant that the canneries could not process the salmon caught quickly enough to avoid the spoiling of the fish. At these times, the processing capacity of the plants could not keep pace with the catching capability of the fishing boats. This led to talk of regulating both the number of boats allowed on the Fraser and the time they were allowed to fish. Anderson received permission to appoint a resident fishery officer at New Westminster and recommended a close time of thirty-six hours each weekend. He also saw a role for the Indian population: "Under careful management they will continue to be, as they now are, of the greatest utility as fishermen, while deriving themselves substantial profit in return for their labour" (Supplement No. 5 to DMF in *SP*, no. 1, 1878, 292). A close time of thirty hours was adopted (from noon Saturday to 6 PM Sunday) both on the Fraser and in the north. However, Anderson modified the hours in 1879 to 6 PM Saturday to midnight Sunday in order to placate missionaries in the north and to allow observance of the Sabbath.

However, Anderson was also clearly aware that the labour of Indian fishermen could be obtained more cheaply than that of white fishermen. Thus an establishment for collecting dog-fish oil on the Queen Charlotte Islands failed "chiefly, as I understand, because White fishermen, at high wages, were employed, while the more economical services of the native fishermen were not utilized" (DMF 1878, 293). In the meantime, a Mr Robertson on Nass River was processing "*oolahan*" for oil and using "natives" to catch the fish, either paying them for the purchase of the fish or wages at $1.25 per day (292). In part, what allowed the labour power of aboriginal fishers to be undervalued in comparison to "white" fishers was their access to the fisheries, which was being threatened by the encroachment of salmon canneries.

In a letter dated 23 October 1877, to A. Smith, Minister of the Department of Marine and Fisheries, E.A. Meredith, deputy minister of the interior, quotes Dr Powell, Indian superintendent of Victoria who notes: "The great impetus given to the establishment of salmon canneries this season excites much talk among the Indians of white people monopolizing their favourite fishing grounds ... I am told that this year, on the Fraser, the cannery firms not satisfied with the extraordinary and almost unprecedented run of fish, have followed them up to the lakes and brought them down by steamer" (DMF 1878, 304–5).

Meanwhile the canneries were accused of wasting the resource. Anderson noted in his report the following year: "Of course, from accidental causes, some sacrifice of fish for mercantile purposes has occasionally happened; but, in such cases the cannery proprietors have usually presented the fish gratuitously to the natives around, who have cured the fish by drying for their own consumption" (DMF 1879, 292). Fish that was spoiled for cannery use was thus given free of charge to aboriginal peoples provided they used it for their own consumption, thus lessening the costs of keeping themselves alive and cheapening the price of their labour power. At the same time, the minister authorized the suspension of fishery enactments as applied to the Indians since, according to Anderson, "the exercise of the aboriginal fishing rights cannot be legally interfered with" (293).

Keeping labour costs low became a preoccupation of salmon canners as they expanded their operations. In 1877 it was estimated that the aggregate expenditure for labour and supplies amounted to over a quarter of a million dollars "and probably exceeds the disbursements of the old established canneries any previous year nearly tenfold" (DMF 1878, 294). And a detailed division of labour was also seen as crucial. "It is of course only by an organized system of action and the minute subdivision of labour that the operations of the industry, from the cutting up of the tin plates, the shaping, the soldering up to the final labelling of the cans, after the insertion and cooking of the contents, can be profitably carried on." Anderson goes on to note that on the Fraser River alone almost 2500 men were employed during the fishing season. While he sees a bright future for "young Indian men" as valuable assistants in the fishery he notes the employment of "a good many Chinese" in the indoor operations whose services "are of special value in the canneries" (DMF 1879, 297). The employment of the Chinese in the canneries was threatened by a law passed in the provincial legislature which caused the cannery proprietors to lodge a protest "and the evil consequences at one time apprehended were averted." Roy (1989, 41) elaborates:

> Only Indians and Chinese would accept the low pay, unpleasant working conditions and uncertainties of the short season. When Indians occasionally complained about Chinese competition, some whites agreed that Indians were equal to the Chinese as labourers and kept money circulating within the country, but the canners claimed the Indians were unreliable and that eliminating the Chinese from the industry would put more than fifteen hundred whites and Indians out of work. Thus, the canners successfully lobbied the province to delay implementation of the 1878 Chinese Tax Bill. The eventual failure of the tax bill permitted the Chinese to remain dominant in the industry until machinery replaced many of them at the turn of the century.

One of the reasons for such poor revenues in 1879 was that the price of salmon in Europe was so low as to be unremunerative. Anderson noted that the Columbia River canners were faced with the same problem. "The market for this article, it may here be mentioned, depends intimately upon the condition of the manufacturing and mining classes in Great Britain and elsewhere, affording to them, as it does, in a convenient form, a very acceptable change from the uniformity of their ordinary diet" (DMF 1879, 280). In his report for 1881, Anderson notes that the price of canned salmon decreased over the years from $16 a case in 1866 to between $5.25 and $5.50 in 1881: "This canned salmon is the poor man's simple food luxury" (DMF 1882, 218). Since the chief consumers were the proletariat of the most industrialized nation in the world, located across an ocean requiring long transport of a heavy and bulky item, it was imperative that costs be kept as low as possible. And labour, or rather its scarcity, became a chief consideration in the take-off decade, the 1880s, of the salmon canning industry. In 1881, the Fraser saw an enormous run of fish with 177,276 cases canned on the river compared to 61,849 the previous year. But Anderson reported that the canneries were underutilized that year due to a "deficiency of labor, arising from the increased demand for railway and other purposes. Hence, too, the cost of putting up the fish was somewhat increased" (202). However, several labour-saving devices were adopted, coming he thought, from the Columbia River, and leading to a saving of 30 percent in the cost of manipulating the cans. He identifies a soldering machine "by means of which cans – placed, after being filled, on a travelling platform worked by an endless chain – are successively presented to the soldering tool, and pass out complete without the intervention of hand labour. There is also a patent apparatus for filling the cans; but this, it is alleged, requires some improvement to render it perfectly effective. It is satisfactory to add that all the machinery connected with such improvements is manufactured within the Province" (202).

Anderson reiterated his claim that the Indian people did not interfere with the fish prized by the canneries since the latter processed the early runs (Suck-Kai, or sockeye) while the later fall varieties of salmon were captured by aboriginal people and dried, these being useless for canning purposes. Anderson represented the dominion government on the joint commission headed by O'Reilly to investigate reserve boundaries. Some regulations were made on the Nass River, where a Mr Gray was deterred from establishing a fishing station in an Indian village since a cannery had already been established adjacent to it. "Indirectly, too, I [Anderson] sought thereby to prevent the probable establishment of future canneries in number beyond the working capacity of the river. It was in view of such circumstances that I framed recently the recommendation of a system of licenses, in order to obtain for the future an effective restraining power" (DMF 1882, 210).

It is clear from the preceding that the department was assuming an increasing role in the regulation of the fisheries as these became ever more commercialized. It would appear from his reports that Anderson was sympathetic to the needs of aboriginal peoples and worried about the encroachment of canneries on their lands. However, his was a paternalistic approach, typical of the more sympathetic bureaucrats, since he never appears to have envisioned a role for aboriginal peoples other than as providers of labour power to capitalist entrepreneurs engaging in the nascent industry. His reports also indicate that he had the best interests of the canning industry as a whole in mind, which caused him to advocate weekly close times and a system of licensing in order to avoid overfishing and overdevelopment in an area where it was very clear that abuses such as these would rapidly and perhaps permanently destroy the fisheries that mattered. While the federal state was regulating the commercial fisheries and involved in determining aboriginal claims, it was very clear that it did so in the interests of general capitalist development. And Anderson's reports stress the important role of aboriginal labour in the prosecution of the salmon canning industry.

Anderson's paternalism is evident in another context. He was also a justice of the peace and reported a "riotous outbreak" at Inverness in 1881. "A body of Chinese, employed in the cannery, were the complainants on the one hand; a European blacksmith, likewise employed there, a counter complainant on the other ... the overbearing conduct of the European complainant had evidently originated the riotous outbreak ... It is obvious that in dealing with large bodies of men, such as these Chinese labourers, and especially in remote localities like this, a certain degree of suavity and forbearance is indispensable ... As a rule these men are easily managed, when kindness and consideration are shown" (DMF 1882, 213).

In the early 1880s, the race and gender of groups immigrating into British Columbia was becoming a primary cause of concern to provincial and federal authorities. Lieutenant Governor Clement F. Cornwall wrote to the Honourable (Acting) Secretary of State, Hector L. Langevin, in a letter dated 10 March 1883, "upon the subject of the necessity which exists of attempting to attract a desirable class of white labor into this Province." Enclosed with it is a report

> Setting out the causes which have hitherto tended to interfere with the immigration of a desirable laboring class to this Province, the consequent introduction of numbers of Chinese to perform indispensable work – the consequent detrimental effect upon the Province of the presence of so large a number of Chinese; and that the true interests of the Province and Dominion alike demand an attempt to ensure the carrying out of the large public works now in hand by means of free white labor so that the vast expenditure thereon may be retained in the country and utilized for its further development. (*SP*, no. 93, 1883, 1)

The fourth point of the report claims that "this semi-slave labor is most prejudicial to the best interests of the country" and recommends requesting $50,000 from the federal government to promote immigration from Great Britain and Europe (*SP*, no. 93, 1883, 2). The request was denied, but an immigration office was established in Victoria, and the federal government proposed to circulate a pamphlet promoting emigration from Great Britain and Europe and offering a $10 bonus to each immigrant over sixteen years of age.

Also enclosed is a petition from the inhabitants of the city of Victoria, signed by its mayor, Noah Shakespeare, and dated 22 July 1882, against the further admission of Chinese immigrants. It was estimated that 20,000 more were likely to enter over the next two years. "This class of immigrants are absorbing the greater portion of the minor industries of the Province, and are able to drive the white labor out of the field" (*SP*, no. 93, 1883, 14). The petition also notes that Chinese do not contribute to revenue since they do not own property! Finally, copy of a report submitted to the executive council of the province, dated 19 August 1882, notes: "The employment of Chinese as carried on is practically establishing a system of slave labor in the Province" (15). It estimates that there were twelve thousand Chinese in the province, over half of whom were employed on construction of the CPR. They are called "a non-assimilating alien race" and reference is made to legislation passed in the United States effective 5th August 1882 restricting further immigration of Chinese labourers, skilled and unskilled, into the country. John Lowe, Secretary of the Department of Agriculture, writes from Ottawa 6 October 1882 that the classes of immigrants most in demand to emigrate to Canada include farmers, gardeners, agricultural and other laborers, and ordinary domestic servants, females in particular. Not needed are professional men, clerks, and persons fitted for "sedentary callings" (18). Clearly the gap identified is a European proletariat, including, most particularly, women.

The result of this concern with the type of person immigrating into British Columbia was the establishment of a Royal Commission on Chinese Immigration in 1885. The next chapter analyses the evidence provided to the commissioners and their views on race and gender. In ending this chapter on the industrialization of the fisheries, however, a word about the nature of corporate control of the industry is in order.

THE DEVELOPMENT OF CORPORATE CONTROL OVER SALMON CANNING

Before the advent of the limited companies in the 1890s, the Fraser River's industrial organization was characterized by low levels of industrial concentration, small firms run by individuals or partners, and by a high incidence of local proprietorship. After the initial investment, the canning proprietors gained fixed capital by re-investing the profits made in the industry. Long-run operating capital, which was especially important to the industry because the salmon market had an eighteen-month cycle from the time the tinplate was ordered until the season's pack was sold, was supplied by commission agents, who made advances in the form of overdrawn accounts on goods in transit. These agents provided canning and fishing supplies and a distribution system to the market as well as capital. (Stacey 1982, 6)

By the mid-1880s, there was overexpansion on both the Columbia and Fraser rivers, and canners began to search for salmon-producing streams in Alaska and northern British Columbia. The Fraser River canning industry consisted of thirteen firms, each tied to a brokerage house (Reid 1981, 323). Victoria was the financial center of the province. From 1871 to 1891, "salmon canning, sawmilling, and the North Pacific seal hunt replaced the mainland fur trade and placer gold mining as the leading staple industries in the Victoria-centred B.C. economy" (McDonald 1981, 370–1). McDonald provides figures that demonstrate that by 1881, salmon canneries and sawmills employed the largest labour forces. Ten years later 85 percent of provincial exports were products of the mines, fisheries, or forests. Fraser River canned salmon was the fastest growing export industry. Between 1876 and 1896 "the value of canned salmon exports shipped to external markets increased five times as fast as the value of forest product exports" (371–2).

Thus salmon canning played a central role in the initial industrialization of the province. It was also linked to the early, still largely mercantile, financial structure centred in Victoria. The next (corporate) stage of capitalism involved the displacement of Victoria by Vancouver, with a corresponding loss of provincial autonomy to national and international finance capital. Salmon canners seized the new opportunities because they saw in them a means for their own financial independence.

The shift in financial control of the industry occurred between 1876 and 1896. Some cannery owners resented the control exercised by commission agents, whose interest differed from those of the canners. One of the industry pioneers, T. Ellis Ladner, noted that agents were less concerned with competition between canners than with financial control over the product and a commission appropriate to their investment. However, the "advent of eastern Canadian banks to British Columbia changed the situation for those canneries not already too involved in the old order. The more independent of the fiscal agent the cannery man was, the more he was able to control his own business.

He could purchase materials at the lowest price and he could finance the introduction of modern plant methods and increase his profits through improved operations" (Ladner 1979, 92).

Central Canadian banks located their regional offices in Vancouver, close to the new transportation terminals. The Bank of Montreal and the Canadian Bank of Commerce became the financial backers of the salmon canning industry. By 1901, canning companies, with the aid of bank financing, could incorporate and thus break the old cyclical dependence on financial agents.

The shift in financial control was paralleled by a shift in transportation from ocean to rail. By the late 1880s transcontinental lines were completed in Canada and the United States. However, while American canners could take advantage of the new means of transportation, Canadian canners faced considerable obstacles. Canadian freight rates, especially on manufactured items moving east, were considerably higher than those in the U.S. Canned salmon packs were a bulky and heavy commodity to move. In addition, the population in the southern and eastern states was large and provided an important market to American canners, especially for the cheaper grades and species of salmon. U.S. markets were closed to Canadian canners, who did not have a large Canadian market. "Prior to 1939, foreign markets, chiefly Commonwealth, absorbed 65 per cent of the B.C. canned salmon production, leaving 35 per cent to be marketed on the Canadian domestic market" (Gladstone 1959, 73). Gladstone also notes that, although Canada is an important producer of fish products, per capita consumption has historically been very low (66). The growth of Vancouver as a railway terminus did, however, stimulate other industries. Canned salmon thus failed to hold its dominant position in the provincial economy, and was eclipsed by forestry products, since agricultural settlement on the prairies generated a demand for construction materials (McDonald 1981, 383).

In summary, from its inception salmon canning in British Columbia has faced fairly weak domestic markets, corresponding dependence on overseas markets dominated by other countries, and sale of a product whose quality is fairly uniform (leading to the need to create loyalty to brand names in order to compete successfully in the market). In spite of these restrictions, however, attempts have been made to dominate the industry.

The first serious bid occurred in April 1891, with the incorporation of the Anglo-British Columbia Packing Company (A.B.C.) in London, England. The primary mover here was Henry Bell-Irving, a native of Scotland who had been a merchant in B.C. since 1883, the company's agent, and chairman of its local committee. He acquired options on nine canneries which he promptly sold to A.B.C., or the "English syndicate." Although control resided in Great Britain, large amounts of capital were subscribed in Canada. One of the reasons for buying canneries was tied to the boat licensing program then in effect on the Fraser River. By acquiring additional canneries, A.B.C. could pool boat licences (each cannery was allowed a specified number of fishing licences) and thus reduce competition from both canners and fishers. Bell-Irving argued:

> My company do [*sic*] not intend this year to work all its canneries because we cannot get enough boats to supply all the canneries with fish – it is proposed to run half the canneries on the Fraser River and use the fish from those boats of canneries not running to put in the other canneries and double up thus reducing expenses, but I think it most essential that there should be a fixed number of licenses to the canners ... so there should be no danger of being frozen out by any combination of fishermen, as canners have money invested and not the fishermen, and if it was not for the canners the fishermen would have a very small market indeed – the local market – and which is a mere nothing to them. (*SP*, no. 10c, 1893, 330)

Acquiring two additional firms, A.B.C. became "the largest producer of sockeye salmon in the world" (Ralston 1965, 25). The merger followed a pattern. In 1889 the British Columbia Canning Company was incorporated in London. Its principals included Fraser River pioneers Findlay, Durham, and Brodie who collectively owned four canneries, three of them located in the north of the province. Another pioneer, Alexander Ewen, also expanded, and by 1889 he owned the largest Fraser River cannery. In 1891 the Victoria Canning Company of British Columbia was incorporated. A major original entrant who remained outside these mergers was J.H. Todd, a Victoria merchant. By the beginning of the 1891 season, there were thus five major groups competing on the Fraser. A.B.C. had bought out all of the American-owned concerns, which by 1881 had represented 30 percent of total fixed capital (Ralston 1981, 300). Both British-backed corporations had local entrepreneurs. Together, A.B.C. and Victoria Canning Company controlled over 60 percent of the Fraser River's sockeye pack. All the companies, except A.B.C., were financially linked to Victoria (Ralston 1965, 19).

This early consolidation of cannery operations was determined to a large extent by the need to buy boat licences in order to secure fish supplies. Attempts at oligopsonistic control stemmed from overcrowded conditions on the Fraser. The number of fishers had risen dramatically, and in the period 1872 to 1888, the number of canneries increased from three to twelve. When the federal fisheries department introduced licence limitation on the Fraser River, the pressure to consolidate increased. In 1889, 1890, and 1891, the total number of licences was limited to 500, with each cannery allotted an average of twenty (Stacey 1982, 13). Both canners and fishers objected to licence limitation (often for opposing reasons), and in 1892 it was abolished.

Predictably, the end of licence limitation led to new entrants, both fishers and canners. Until the turn of the century (and to a more limited degree until the end of the First World War), it was fairly easy to enter the business of canning salmon, because technological development was minimal, the operations themselves were labour intensive, and labour was cheap. Gregory and Barnes (1939, 30) describe the situation prevailing throughout the Pacific northwest: "The greater number of canneries prior to 1893 was owned by single proprietors or partnerships ... Except for the more elaborate ones they could readily be moved from one site to another to adjust to changing fishing conditions and competition. They were devoid of much machinery; their costs were low, and most of the early packing and handling prior to 1903 was done by hand ... Often the canneries were enlargements of salteries that preceded them."

On both sides of the border, there was a trend toward a fish pack divided between a small number of large companies who dominated the industry and a large number of small firms. While A.B.C. initially acquired some advantage, within ten years the situation had come full circle. "By 1901 the level of concentration had reverted to its pre-merger position" (Reid 1981, 320).

With renewed competition came a new bid for control of the industry. However, the pattern differed this time because, with the availability of bank capital, canners could assume greater control over their operations. Three of the largest operators (Victoria Canning Company, Alexander Ewen, and George Wilson) joined forces with Aemilius Jarvis and Henry Doyle. Jarvis had financial connections with central Canadian financiers, going back to 1892, when he had established Aemilius Jarvis and Company, Investment Bankers. Doyle, on the other hand, had detailed knowledge of the salmon canning industry, going back to his time with Doyle Fishing Supply Company of San Francisco. Doyle had an added advantage in that he had conducted most of his business in the United States and thus appeared to be a neutral partner to provincial canners, that is, without an interest in specific plants. Financial backing would be provided on condition that the principals obtain control of 60 percent of the operating plants.

Canners who sold out to the new company, British Columbia Packers, were asked for a commitment to not participate in the industry for at least seven years. J.H. Todd originally gave his active support to Doyle and the new merger. However, he pulled out of the agreement after most of the plants had been purchased. With much of the competition eliminated, Todd stood to gain substantially from the new balance of power (Reid 1981, 315–19).

Doyle was familiar with two similar mergers in the United States. In 1893, the Alaska Packers Association was formed, following a merger the previous year when 90 percent of the producers combined their operations. In 1889 another combine emerged, the Columbia River Packers Association. In British Columbia merger was facilitated in 1901 when there was a heavy pack carryover and many canners became indebted to the banks. The Bank of Montreal held half the salmon canners' accounts, and the Canadian Bank of Commerce held another 40 percent (the remainder was with Molson's Bank which, in 1942, was taken over by the Bank of Montreal). In 1902 the three banks approved the proposed amalgamation. Jarvis had already formed a syndicate and was acquiring subscriptions from central Canadian businessmen. The new company received its formal charter 8 April 1902 in New Jersey, and was called The British Columbia Packers' Association of New Jersey (Lyons 1969, 230–3).

In the United States, "institutional barriers" and lack of "widespread credit markets" hampered the centralization of capital. New Jersey law and a narrow interpretation of the Sherman Antitrust Act partially overcame legal barriers. Then, between 1898 and 1902, there was a large merger wave which "drastically transformed the structure of large business in the United States" (Edwards 1979, 42–3). The mergers in Alaska, British Columbia, and on the Columbia River were part of the emerging pattern of capital concentration and the birth of the modern corporation.

In the 1890s new low-cost salmon producing areas emerged in Alaska and on Puget Sound in the United States. Puget Sound production intercepted runs headed for the Fraser. Reid (1981, 326) estimates that while, in 1890, 97 percent of Fraser fish was canned on the river, by 1900 the proportion had fallen below 40 percent. With lowered Fraser River production, ownership of northern plants became very important. "[The] northern canneries had shown historically much larger profits per case than those on the Fraser. Any company, therefore, wishing to control the Fraser River fishery would be financially stronger if it also possessed northern plants" (Sinclair in Reid 1973, ii).

When it was formed B.C. Packers took possession of twenty-nine of forty-eight canneries on the Fraser and twelve more in the north. By 1902 it had closed one-third of pre-existing canneries and had doubled remaining plant capacity by using machinery and equipment from idle plants. Until that time, salmon canning had been an assembly-line, manually intensive process. After the turn of the century, machines were introduced at various points along the assembly line. Mechanized lines in turn increased the amount of capital required to enter canning, and allowed several lines to be combined in one plant. The multi-line cannery was superior to the single-line operation. Production did not have to stop in order to change to a different size of can, and surplus buildings in idled plants proved ideal for storage. Stacey (1982, 19) estimates that by 1905, fifteen plants nearly equalled the capacity of the twenty-nine canneries purchased three years earlier on the Fraser.

In the twentieth century the two largest and most dominant corporations in the fishing industry have been B.C. Packers and Canadian Fishing Company, Ltd (Canfisco). Canfisco was originally a halibut company bought in 1909 by a U.S. firm, New England Fish. In 1918 Canfisco entered salmon canning with the purchase of Home Plant, situated on the Vancouver wharf adjacent to the city's Japan and China towns. The two corporations became major rivals. Several medium-sized companies, like J.H. Todd & Sons Ltd, also operated salmon canneries at strategic locations along the coast. Nelson Brothers Fisheries, incorporated in 1929, was another medium-sized rival. The two brothers began in the industry as fishers in 1919, trolling on the west coast of Vancouver Island. In 1933 they purchased their first salmon cannery, St Mungo, on the Fraser, and subsequently purchased canneries in other locations. In 1940 they began operations in Prince Rupert, and in 1943 they bought Port Edward, located nearby, from B.C. Packers. In 1955 Nelson closed St Mungo and built Paramount Plant in Steveston, on the Fraser (Lyons 1975, 405, 459, 522). In the latter part of the 1950s, Canfisco and B.C. Packers jointly took over J.H. Todd & Sons. A decade later, Nelson Brothers suffered a similar fate. Both sets of transactions were kept highly secret. By the 1980s B.C. Packers was operating the two largest salmon canneries in the province. One is located in Steveston, Imperial plant, of which Paramount is part, and Oceanside plant in Prince Rupert. Oceanside was bought from Canfisco, rebuilt and enlarged, and renamed Prince Rupert Plant. Until that acquisition, Port Edward had been B.C. Packers' major northern operation.

Toward the end of the 1970s, New England Fish began to experience severe financial problems, although its Canadian subsidiary was solvent. Meanwhile, in 1962 George Weston Ltd, a central Canadian food conglomerate, took control of B.C. Packers. In 1980 Weston bought the majority of Canfisco's northern operations (UFAWU 1984, 13). Canfisco continued to operate Home Plant in Vancouver, although its scale of operation was severely reduced, leaving B.C. Packers in the dominant position. Its major competitor in the north is a fishers' cooperative, the Prince Rupert Fishermen's Co-op, which began canning salmon in the early 1950s, and expanded into processing other fishery products in subsequent decades, upgrading and modernizing its facilities on the Prince Rupert dock.

While salmon canning was the initial vehicle for controlling the industry, after the turn of the century other fisheries became important, particularly with the introduction of refrigeration on fishing boats, in processing plants, and for transportation to markets. Thus, just when salmon canning began to be mechanized, making it difficult for new entrants, other technologies and fisheries facilitated competition by other means (Pinkerton 1987a, 66–91). However, the companies who originally wrested control of salmon canning retained control of the entire industry. Reducing competition by buying up salmon canneries in strategic locations, they gradually, over the course of the twentieth century, concentrated facilities in urban areas (when the technology was available to permit transporting fish over longer distances). Correspondingly, they enlarged the remaining salmon canneries, converting them into multi-purpose plants, with freezers, reduction plants, fresh fish operations, salmon and herring roe plants, and canning lines for crab, clams, and other fishery products. Maps 2 to 4 show this trend.

A new type of foreign investment began in the 1970s, centred around the market for herring roe. The Japanese, having overfished their own stocks, searched for new areas, resulting in a new provincial fishery, since to that time the market for herring had been limited to its use as fertilizer (reduction). Japanese capital flowed into the roe herring (and roe salmon) fisheries and into the fresh/frozen markets. As will be shown in chapter 5, prosecution of this new fishery had different impacts on fishers and plant workers and provoked an industry-wide strike.

While those companies who early dominated salmon canning retained control when new fisheries were exploited, there was a parallel in the historical development of shore work. New fisheries and new products were frequently accommodated into existing patterns of work within the canneries. For example, the herring roe season precedes salmon runs. In addition, herring roe can be frozen and its processing spaced out to prolong cannery work (thus ensuring enough weeks for workers to enable them to collect unemployment insurance during the winter months).

To summarize briefly, until the turn of the century salmon canning was the major form of capitalist production in the B.C. fishing industry. Afterwards, canneries were expanded to accommodate other processing technologies and fisheries. While competition in the nineteenth century took place in salmon canning, with technology competition became an expensive venture. It became far easier to compete by means of other technologies (like supplying fresh/frozen fish markets) and/or other fisheries (like halibut and roe herring). As chapter 5 will show, there were serious implications for unionized shoreworkers from competition from these small, largely unorganized plants.

CONCLUSION

In this chapter I have tried to demonstrate how the social construction of race and gender is central to capitalist appropriation of property rights (aided by the state) and utilization of labour power, using the First Nations and the fisheries of British Columbia as examples. It is difficult to demonstrate the theoretical formulation because of the linearity of the printed word which forces one to separate interlocking strands of relationships that are complex, multi-layered, and ever changing. In the next chapter I continue the analysis by incorporating yet another racial construction: that of the "Chinaman" and the "Jap." Unlike the "Indian," a category that emerged under colonial rule, these categories were a product of British imperial expansion which resulted in the release (rather than proletarianization involving a transformation of a pre-existing mode of production) of cheap wage labour from countries with state structures that became major political forces on the world stage. Thus, unlike "Indians" who constituted little colonies within Canada, the employment of Chinese shoreworkers and Japanese fishers was conditioned by international imperial relations of shifting balances of power. In other words, the decisions of provincial politicians to ban further Chinese immigration, and laws restricting the movement of the Chinese within the province and their ability to engage in enterprises were contingent on the federal government's dependence on British imperial policy. When China and/or Japan were wartime allies, the federal courts most often struck down racist legislation. Thus another dynamic was brought into play involving the cheap wage labour of Chinese and Japanese labourers in the fisheries that had connections to, but was also often in direct conflict with, utilization by salmon canners of the labour power of aboriginal peoples.

4 The Dialectics of Cheap Wage Labour

ABORIGINAL CANNERY WORKERS

By the 1880s most coastal villages included salmon canneries as part of their seasonal migration. The Fraser River canneries, from their inception in the 1870s, almost guaranteed employment to the Coast Salish living in the vicinity, while the Tsimshian were secured places in the Skeena and Nass River plants. But soon the entire coast was involved, including the tribes living on Vancouver Island and along the central coast. W.H. Lomas, Indian agent for Cowichan Agency on Vancouver Island noted that in the summer of 1881, several villages in the southern part of his district were almost deserted as men, women, and children found employment on the Fraser River. He estimated they would return with over $15,000 in wages (Canada 1872–1902, Department of Indian Affairs [hereafter referred to as DIA] 1882, 160). Meanwhile, P. McTiernan, Indian agent for Fraser Agency, moaned that the Indians went off to commercial fisheries, neglecting the cultivation of reserve land. "There is no class of labourers to compete with them at the fisheries or at steamboating on the Fraser River. Their women, also, who are very industrious, are profitably employed at the fisheries during the fishing season, making nets and cleaning fish for the canneries ... The Indians love working in batches together, and much prefer the above kind of employment to agricultural labour" (166).

The following year, 1882, 1300 aboriginal men were employed on the Fraser River, earning an average daily wage of $1.75 for a season lasting approximately ninety days. Most of them fished. Four hundred aboriginal women were employed cleaning and canning salmon for $1.00 a day (DIA 1882, 61). While such a short season could not provide sufficient employment to feed a European family, the pooled wages for a whole village represented a considerable amount of money. The canners required a large number of labourers to process all the fish (refrigeration techniques at this time were almost nonexistent, and fish had to be immediately processed or it would spoil). Catching capacity tended to outstrip canning capacity, and fishers were sometimes limited in the number of fish they were allowed to land in a twenty-four-hour period. Indian contractors were sent to the coastal villages to persuade everyone to relocate at the canneries for the season. There were problems holding the labour supply, especially in these years of railroad construction and the availability of wage labour in a thinly populated province (Knight 1978); in 1883 the canners caused much discontent among their employees when they held wages until the second run ended.

When the first run of salmon is over on the Fraser River, the Indians are told by the managers or owners of the fisheries, that they have no more work for them until the second run commences, which often is a delay of two weeks; they retain the Indians' money as security that they may not go home or engage in any other occupation until they want them again, therefore, the Indians are obliged to remain idle about New Westminster for that length of time or forfeit their wages. Some Indians come hundreds of miles to labour at the fisheries, and to have them subjected to such unfair treatment is certainly a great grievance and one they bitterly complain of. (DIA 1883, 46)

Locations near European settlements also caused incomes to be diverted from buying commodities useful to the village to the cultivation of individual habits like gambling and alcoholism. A booming business in New Westminster, Victoria, and on Puget Sound catered to these activities. In addition, aboriginal women were recruited as prostitutes in these centres, which contained a heavy concentration of single European men. One of the chief reasons Indian agents sought to keep the Indians on the reserves was to keep them away from the cities and their attendant vices.

Gradually, salmon canners found a more subordinated labour force among Chinese male labourers. In 1884 there was less work available due to increased employment of Chinese men inside the canneries.

The Indians from all parts of this agency [lower Fraser] complain very much this spring and summer of how they are undermined in the labour market by

Chinamen, especially in all kinds of light work, where the Indian women and their boys and girls used to be employed. Although many of them have came [*sic*] long distances this season to the fish cannaries [*sic*], very few of them got employment, as their places had been taken by Chinamen, in cleaning and canning the fish; they are also doing all the washing and ironing in private families, what Indian women used to do heretofore. The poor Indian women and old men, and their boys and girls, used to make considerable money every summer picking berries and selling them to white people. This summer large numbers of Chinamen went into that business too, and almost completely ruined the Indians. (P. McTiernan, Indian Agent, Lower Fraser Agency, 15 August 1884, DIA 1884, 104)

Chinese men displaced aboriginal women in cannery work as well as taking their place in domestic work for private families, doing washing and ironing. The canners in turn had to compete with railroad construction, which in the mid-1880s imported large numbers of contracted Chinese labourers. In these years, therefore, large numbers of aboriginal women were hired to work in the Fraser River canneries, but that situation was temporary. Fraser River canneries were close to urban centres, in which Chinese contractors operated and which also contained growing China towns with abundant Chinese labour. In the north, aboriginal peoples continued to be hired in large numbers because they lived in the area and it was extremely costly to transport labourers from the southern part of the province (84).

The sheer number of aboriginal peoples congregating at New Westminster, on the Fraser River, each spring created problems. In 1885, three thousand came from all over the coast seeking cannery employment and camping along both banks of the river, from New Westminster to the mouth of the Fraser. Indian agents were not blind to the potentials of such a large gathering. "Indian affairs require careful handling, as, although tribal feuds and jealousies have for long kept distant bands from uniting, still the present labour fields throw the different bands together, and they hear each others grievances, and although a feeling of discontent is not likely to make any uprising on the land question possible, still it is this feeling which encourages those murders of isolated miners and settlers which were so common a few years ago." W.H. Lomas, Indian Agent for Cowichan Agency, also expressed his views about the character of aboriginal peoples, what made them different from Europeans. "In my opinion the chief cause of the failure of so many different missionary efforts for the advancement of the native races is the mistake which all seem to make, of judging and treating Indians by European standards, forgetting that though the Indian is a close reasoner, his character is a mixture of child-like suspicion, credulity and selfishness, but with a keen sense of humor" (DIA 1885, 81). In his report (dated 7 August 1885), Lomas referred in particular to the "Metlakatlah land troubles" and the North-West rebellion.

This comment indicates one reason for the preference in hiring Chinese instead of aboriginal labourers. The aboriginal population was dispersed over a large area and still controlled, in a limited way, its own means of subsistence. Village and community organization was strong. Most aboriginal labourers camped near the canneries in village groupings, and often worked alongside village members. In other words, they were only a partially subordinated labour force, unlike Chinese labourers, organized under the Chinese contract system.

Chinese contractors, or middlemen, supplied, supervised, and provisioned the bulk of the labour force for each cannery. Their labour forces were hired for the entire season, while aboriginal labourers working in the factories (that is, the women and children) were only given work when it was available. Race and gender distinctions were very important. Although aboriginal women in northern canneries might actually do work Chinese men performed in the south, within any one cannery there was a strict gender and racial division between the work performed by Chinese men and aboriginal women. The women, and their children, received the lowest wages and least secure conditions of employment.

Geographical distance served to separate Chinese labourers from pre-capitalist relations of production. A similar method could not be employed for aboriginal labourers, nor was it desirable since salmon canners, especially in remote areas, required proximate labour supplies (thus cutting costs of transporting labourers). As discussed above, the potlatch was the institution binding the socioeconomic structure of the First Nations. Indian agents and missionaries tried to stop the ceremony for years, with varying degrees of success. In 1885 the potlatch was legally banned by special amendment to the Indian Act. Capitalist penetration of aboriginal economies was probably more successful than legal prohibition, but it does indicate a degree of awareness of its significance on the part of the European population. Especially among the Kwagiulth, the practice continued in secret, and Indian agents complained bitterly of its persistence. The Tsimshian also resisted. They refused to allow any Indian agent to come and reside with them, and the Indian Act could not be enforced in the north for many years. An agent was only appointed in 1887, and he established his residence in the mission village of Metlakatla. The Tsimshian also prevented surveyors from establishing boundaries. Armed cruisers were sent from Victoria, and Indian leaders were arrested (DIA 1887, x–xi).

Northern aboriginal groups could resist more effectively because European settlement did not yet extend that far, and contact with Europeans was minimal. In the south, however, urban growth made subsistence increasingly difficult. For example, in 1886, the Fraser River salmon run failed. The First Nations were caught between modes of production: they could not acquire the fish through traditional methods since the fish was absent, and they could not supplement with food commodities because they could not earn the cash income.

In European society, sexual inequality structured labour in such a way that a different value was assigned to the labour power of each gender (and the labour of women was often granted no value whatsoever). This allowed industrial capitalists to employ female labour power below its costs of production and reproduction. In British Columbia, salmon canners were primarily of European extraction, and they brought with them an ideology that valued the labour power of men and women differently, as well as that of non-European races. While aboriginal labourers, male and female, were paid lower wages than European male labourers, aboriginal women (and children) were paid the lowest wages. While gender and racial characteristics have nothing to do with the operation of capitalist relations of production in the abstract, they have everything to do with the way those relations are practised. That is, structures of inequality based in patriarchy are used to structure labour forces, paying labourers as low a wage as possible. These structures of inequality allow large groups of labourers to be paid below the costs necessary to produce and reproduce their labour power. While gender and racial divisions are introduced in the work place, pre-capitalist relations of production are necessary to allow these labourers to survive. Thus class relations interact with patriarchal structures, and these in turn interact with existing economic relations to bring potential labour power into the capitalist sphere of production and market exchange.

Women played a crucial role in the maintenance of aboriginal economies, and this role appears to have been strengthened. An Indian agent for Cowichan district observed: "The Indian women and children are always the most eager to go to the hop fields, where they always earn considerable sums of money, and, amongst these Indians, the wife's purse is generally entirely separate from the husband's" (DIA in *SP*, no. 6, 1887, 92). Women used wages to buy clothing, furniture, stoves, and sewing machines, as well as staples like flour, tea, and sugar. Many young aboriginal men spent their incomes in the urban centres, causing people on the reserves to rely on the earnings of the women and children. Indian agent W.H. Lomas, Cowichan Agency, provided an insightful summary of the effects of wage labour on village life:

All the younger men can find employment on farms or at the sawmills and canneries, and many families are about leaving for the hop fields of Washington Territory; but the very old people who formerly lived entirely on fish, berries and roots, suffer a good deal of hardship through the settling up of the country. The lands that once yielded berries and roots are now fenced and cultivated, and even on the hills the sheep have destroyed them. Then again, the game laws restrict the time for the killing of deer and grouse, and the fishery regulations interfere with their old methods of taking salmon and trout. With the younger men the loss of these kinds of food is more than compensated for by the good wages they earn, which supplement what they produce on their allotments; but this mode of life does away with their old customs of laying in a supply of dried meat, fish and berries for winter use, and thus the old people again suffer, for *Indians are often generous with the food they have taken in the chase, but begrudge giving what they have paid money for, without suitable return.* (DIA 1888, 105, emphasis added)

The quote illustrates the dilemma of a people caught in two different economic sets of relations. By the end of the 1880s a pattern had emerged. In the early spring the men went hunting and trapping, while the women and old men prepared gardens, planted potatoes, etc. However, as Lomas indicates, the yield from these sources had shrunk. Thus, when the salmon canneries opened in May or June, the entire village, except for the very old, was deserted as people sought to supplement their food supplies with food bought with wages. After the canning season, many families travelled to the hop fields where they could extend their wages (especially women, who did not have as many employment opportunities as did young men). After the canning season, other groups travelled to their fishing stations to catch and dry fish for their own use. On reserves with land unsuitable for farming, people purchased potatoes and vegetables from others, and concentrated on manufacturing activities during the winter months, including building boats, canoes, and household furniture. Women able to afford sewing machines made clothes for their families and to trade or sell (knitting machines were also used to make, for example, the famous Cowichan sweaters).

However, the availability of wage employment was partially caused by the small size of the urban population. In the 1880s the situation began to change. Ironically, salmon runs at the end of the decade were phenomenally large, but the employment of aboriginal peoples declined in proportion to the total labour force employed. In 1894 Superintendent A.W. Vowell concluded:

It is noticeable that within the last few years there has been a falling off in the gross earnings of the natives in British Columbia, which may be accounted for by the gradual influx of settlers of every nationality into the province, which increases each year. The Indians do not now, nor can they expect to in the future, make as much money as formerly in any line of industry or enterprise where the natives used to be the only people available for such employment and pursuits; whitemen [*sic*] and Japanese, and others, are at the present time to be seen in all directions and in great numbers competing with them in the labour market, and in the occupations of fishing, trapping and hunting, etc. (DIA 1894, 202)

The trend Vowell notes was partially suspended in the fishing industry over the next few years because the enormous supply of fish led to the construction of new canneries. Aboriginal peoples were employed in large numbers, but their proportions lessened, especially on the Fraser. Contractors could offer cheap Chinese and Japanese labour (Japanese men competed with aboriginal and European fishers). Canners kept jobs segregated by race and gender, so that encroachments made by specific groups of labourers were perceived in racial terms rather than as a method allowing employers to keep labour costs minimal.

As table 4 indicates, between 1885 and 1890 the number of canneries increased from nine to thirty-two, most located on Fraser River. Fishery officers feared overfishing would exhaust salmon stocks, as had occurred on the Columbia River system. In 1892 a fishery commission was appointed to investigate the problem. The evidence submitted gives some indication of the size and composition of cannery labour forces, although the evidence varies considerably. Overall, the average Fraser River plant employed approximately 8 white men (foremen, firemen, and watchmen in charge of the retorts), 100 Chinese men, 40 to 50 Indian women, and 18 to 20 boys (Indian and Chinese). One canner paid Indian women a dollar a day while white boys received two dollars. Chinese men were generally paid monthly by their contractors. In Wadham's cannery, for example, the "boss Chinaman" received 50 to 70 cents a case or a little over one cent per can. In this plant, the Chinese contractors hired Indian women "and of course these Chinamen pay the Klootchmen." Lord goes on to note, "It would not pay any white to do the work the Chinamen do for the pay, or anything like what the canneries would be willing to pay ... a white man would starve to death." Indian women received 10 cents an hour (12 1/2 cents at Wadham's), while Chinese men received between 30 and 45 dollars a month. White salmon (which, unlike the red fish, had no value in the can) was given free to the Indians (testimony of F.L. Lord, BCFC, 1892, 178–9).

Numbers employed in any one day, however, varied a great deal. The canners retained a core of white and Chinese men to whom they guaranteed steady employment (white men were hired directly onto company payrolls and paid by the canners). Around this core, the canneries required a number of casual labourers who could be called in at any time, and who might have to work around the clock. This was the reason it was so handy to have aboriginal families camped near the plants. In fact, one pioneer canner, Alexander Ewen (BCFC 1892, 117), claimed it was necessary for canners to have licensed boats in order to offer employment to Indian men to fish, thereby making sure they brought their families to the cannery. "The real reason that you want to have those boats of your own and get Indian fishermen as they bring their families around and you have Indian women and boys, and some of the men, not fishermen, to work in the canneries, and when this extra fishing comes on you can take off your own boats and get off to work in the cannery." Later in his testimony Ewen made a very prophetic observation. When asked if he thought the canneries "a benefit to the Indian population" he replied: "Well, I believe they are; it is work the Indian naturally likes to work at and they get good wages and whole families work at it. If the Indians departed from working at that, I think there would be nothing for them but to go back to the reservations and let the Government feed them" (121). Marshall M. English, representing A.B.C., employed Indians: "Because we have to – we cannot import labor from the east and employ them for one or two months only. These people come from all parts of the country and bring their belongings and fish for five or six weeks and then go home again – white people would not do this" (142). Table 5 reproduces the names of canneries operated in 1892, with the names of their owners, year first operated, number of boats attached to each cannery, number of employees and pack sizes (number of cases of canned salmon per season; pack sizes vary considerably from season to season).

The primary concern of the B.C. Fisheries Commission was the question of the disposal of fish offal. Canneries were generally built on pilings out over the water. Testimony of John Ross, "a resident of New Westminster for sixteen years, a native of Great Britain and a fisherman for sixteen years" (BCFC 1892, 151), indicated that the head, tail, and entrails of millions of fish, during a concentrated period of time at the height of the summer months, were pushed off the table upon which they were cut "into a hole and if there is no boat underneath it falls into the river" (152). Residents living along the Fraser River used it as their source of drinking water, and cases of typhoid fever were reported. Several experiments to convert the offal into oil had been tried but were not successful – that is, not profitable. One of the chief difficulties was finding suitable markets.

Table 5
Canneries Operated in British Columbia, 1892

Owner or Agent	*Name of Cannery*	*First Year Operated*	*No. of Boats*	*No. of Hands*
FRASER RIVER				
Bon Accord Fishing Co.	Bon Accord	1879	27	211
	Sea Island	1890	40	240
J.H. Todd & Son	Beaver	1889	35	270
	Richmond	1882	35	270
Ewen & Co.	Ewen's	1876	40	332
H.E. Harlock & Co.	Harlock's	1882	30	220
B.C. Canning Co. (London)	Fraser River Cannery	1876	30	270
Victoria Canning Co.	Delta	1887	40	250
(T.E. Ladner, Mgr)	Wellington	1880	40	230
(4 canneries)	Laidlaw's	1878	40	250
	Holly	1890	40	250
ABC Packing Co. (Vancouver)	Wadham's	1887	40	3000
H. Bell-Irving, Agent	British Columbia	1887	36	in
(8 canneries)	British American	1887	36	total
	Canoe Pass	1889	36	for
	Phoenix	1887	36	ABC
	Gary Point	1889	36	
	Annandale	1891	36	
	Dumfries	1891	36	
Terra Nova Packing Co.	Terra Nova	1892	35	270
Lulu Island Canning Co.	Lulu Island	1893	30	210
Pacific Coast Packing Co.	Pacific Coast	1893	30	290
Stemston Canning Co.	Stemston	1893	30	270
Short & Squires	Imperial	1893	30	290
Canada Pacific Packing Co	Canada Pacific	1893	35	400
Brunswick Canning Co.	Brunswick	1893	30	262
SKEENA RIVER				
Rithet & Co.	Standard	1890	40	185
Byrnes & Cuthbert	Balmoral	1886	30	182
Dalby & Claxton	Royal Canadian	1892	40	242
Cunningham & Son	Skeena	1883	39	220
ABC Packing Co.	British American	1883	30	196
	North Pacific	1889	40	153
B.C. Canning Co. (London)	Windsor	1878	40	209
Turner, Beeton & Co.	Inverness	1878	40	185
Cunningham & Rood	Low's Inlet Cannery	1890	8	136
NASS RIVER				
Federation Canning Co.	Federation	1888	40	169
Rithet & Co.	Cascade	1889	30	141
B.C. Canning Co. (London)	B.C. Cannery	1889	30	157

Table 5 (cont'd)
Canneries Operated in British Columbia, 1892

Owner or Agent	*Name of Cannery*	*First Year Operated*	*No. of Boats*	*No. of Hands*
RIVERS INLET				
B.C. Canning Co. (London)	Rivers Inlet Cannery	1882	35	200
	Victoria	1882	35	200
McNeil & McDowell	Warnock	1884	35	186
S.A. Spencer	Alert Bay Cannery	1881	8	100
H. Price & Co.	Price's Cannery	1836*	24	73
Dearny & Skitbolt	Nanaimo Cannery	1893	8	62

Source: Annual Report of the Department of Marine and Fisheries, in *Sessional Papers*, 1894, no. 9, no. 11, p. 286.

* This would appear to be an error.

The other issue looming large was licence limitation: who should receive licences, what proportion should go to canners, fishers, freezers, and settlers, and whether licences should be granted only "to resident British subjects." Non-British Europeans (for example, Italians and Greeks) had entered the industry as fishers. One fisher, John Kelly, a native of Newfoundland who had been in B.C. for two years, complained that he was unable to obtain a licence. He had come and bought a house but found it difficult to support his family; his two sons also wanted to engage in fishing. "I tell you gentlemen it is a shame; here are Italians and Chinese and all sorts at work fishing, and good Englishmen and British subjects on their own soil cannot get a license" (BCFC 1892, 381–2). Kelly's testimony was greeted with applause (although it was made clear in the overall testimony that the Chinese did not fish for the canneries). When asked if he thought that Italians and others were as good as Englishmen, he replied an emphatic no (they were not as good). John Ross, a native of Great Britain, resident of New Westminster for sixteen years and sixteen years a fisherman, noted that the Chinese did not fish. Upon further questioning he affirmed that Indians and "whitemen" fished – all nationalities, meaning "Greeks, Italians, Chilians [*sic*], Sandwich Islanders, etc." Commissioner Armstrong categorized this group as "outside foreigners" (153). There was a very clear distinction being made here along lines of "nationality," with arguments put forward by fishers that licences be granted to "British subjects," thus excluding newly arrived immigrants from Europe outside Great Britain. We see here the complicating factor of ethnicity added to concepts of race. Ethnic differences have been especially important in the ideology of nationalism and in the emergence of the nation state. In the establishment of the province of British Columbia, ethnicity was also an important factor alongside race. Thus, a "white" British Columbia was conceptualized as "British."

The testimony also reflected a clear division of labour within the canneries by race and gender; that is, white men, Indian women (referred to in a very derogatory way as "Klootchmen" or "Klootchies") with children often included, and "Chinamen." Canner F.L. Lord, native of the U.S. and B.C. resident for fifteen years, described the labour process in some detail. About 100 people were employed in Wadham's cannery. When asked how many white men were employed in specific positions, he replied foreman of the cooking, fireman, and several other assistants – a white inspecting the filling and white men in charge of retorts, timing and keeping the proper temperature – and watchmen, etc. (BCFC 1892, 178). (See the appendix; the classifications listed there reflect the expansion of this group of positions, assigned to white men, in the 1940s, as mechanization required more of this type of personnel.) Lord estimated he employed about eight white men with the rest "Klootchmen and Chinamen," with Chinese predominating. "The Indian women wash the fish, and pile the cans away, and such work as they can do." He noted that in canneries using the contract system, Indian women were paid from the contract. "Q. Oh, then the whole of the work is done by the Chinaman boss? – A. Yes, the whole thing; and of course these Chinamen pay the Klootchies." Indian women were generally paid 10 cents per hour while Chinese labourers earned $30, $32, or $45 per month, although very few received the highest wage. Lord also noted that the Chinese and Indian labourers received all the fish they wanted "gratis" for their own consumption, although he seemed to be referring to the species unsuitable for canning since he noted that all the white salmon was given to the Indians for food consumption (178–9). As previously mentioned, during these years annual reports on the B.C. fisheries included estimates not only of the food fishery for Indian consumption (valued at $3 million for 1893) but also of the home consumption of Chinese labour, valued at $150 thousand for 1893. Although these were large sums, there was no indication as to how the estimates were calculated.

The testimony of fishers makes clear that salmon canners were not the only ones dissatisfied with licence limitation on the Fraser. With the enormous expansion of people, shoreworkers and fishers, working in the industry, race and cheap labour became hotly disputed issues. Just as salmon canners made use of contract labour inside the canneries, they also tried to institute a similar strategy on the fishing grounds as a means of ensuring a cheap supply of fish to the plants. But fishing, unlike cannery work, appealed to many men of European background.

James Wise of New Westminster identified himself as one of the first fishers on the river; four years before he testified to the commission, he had sold out to Ewen. He noted: "There is nothing in any part of Canada or the States where a monopoly is given to the few like here ... Many have come with their families but they could not get a license and then they are under the thumb of men who give small pay, for they get Chinamen and Indians next to nothing" (BCFC 1892, 5–6). John McLashlan, native of Scotland, was unable to obtain a licence and thus worked for a cannery as a net-man. When he was asked about the difference between the labour of the Chinese and white men, he replied: "Well, the white men do nothing as regards the fish. The white men look after the Chinamen and have the higher classes of work." He received $40 per month until the sockeye run began and $60 per month afterwards. He noted that some of the Chinese labourers were paid by the piece and some by the day (24). This testimony indicates the extreme variation in how various groups of workers were paid, variations also among individual plants and geographic locations. What emerges is a hierarchy of wage scales with white men at the top and Indian women and children at the bottom. It was only with unionization that some uniformity in wages began to be introduced, but the hierarchies remained in place, as can be seen in the appendix and as is discussed in the next chapter.

The result of the commission hearings was the abolishing of licenced boats on the Fraser River. The end of the decade witnessed a series of phenomenal runs to the river. In 1896, for the first time, a Pacific fish – salmon – generated the highest commercial revenues in the Canadian fisheries. By 1901, seventy-seven canneries were operating in the province, and thousands of fishers of all nationalities earned their livelihood on the Fraser alone. This was the golden age of the provincial salmon canning industry. During these years, all available labour, including aboriginal, Chinese, Japanese, and European, was in demand, although job structure continued to mirror gender and racial divisions.

As indicated in the previous chapter, at the turn of the century there was an attempt to gain corporate control of salmon canning. When B.C. Packers formed, the corporation closed a number of the canneries it had bought, and assumed a dominant position in the industry. The two world wars served to erode that position, however, by creating a high demand for canned salmon in countries allied to Canada (as a source of cheap protein that was easily preserved). New entrants flocked into the industry to take advantage of the potential high profits. After the Second World War, Superintendent Vowell's prediction was finally and irrevocably borne out. One after another, salmon canneries operating for the better part of the century began to cease operations. The First Nations in the north suffered especially severely, since that part of the province was settled at a later stage than the south. Canning operations were increasingly combined and moved near large urban centres (Vancouver and Prince Rupert). Canners could now recruit women from newly immigrating overseas populations (Japanese, East Indian, Portuguese, to name the most important) in addition to aboriginal women. Maps 2 to 4 show this trend.

By the 1950s, aboriginal women could boast of three generations of women in their families who had worked in the same canneries. When the plants closed, the most mobile and best workers relocated to canneries still operating. But most were forced back on reserves with no paid employment. The state filled the gap with unemployment insurance, welfare, and old age pensions. It is fitting to conclude this section with the testimony of aboriginal women cannery workers.

Kinship could be traced according to which women worked in which cannery. Mary Hopkins was a retired cannery worker.

> I don't know when I was born – eighteen hundred and something; it's in the papers there. I am Bella Bella; my mother is from there. They all worked in canneries, my mother, my grandmother. I started when I was sixteen. Bella Bella, Rivers Inlet, later when I got married Butedale, Klemtu. Oh, I liked it. I really enjoyed working in that cannery here. We used to hand fillet the herring. Every summer I worked, worked long hours. Lots of fish; start at eight in the morning, no time to rest. We got lots of money then. All the women were working.
>
> When the canneries were closed, there is [*sic*] no more jobs for us. All the women have time. We were really sad when we heard it; some of them cried. Now we only get welfare. I get old-age pension. (Steltzer and Kerr 1982, 47)

This woman judged her wages to be good. Since filleting is the most highly skilled job for women in fish plants, it commanded higher wages than, for example, washing fish. Hopkins also noted that all the women worked, resulting in a relatively good income for the village. And working twenty hours in a row when one is paid by the hour, even without overtime pay, would result in a high money income for that period of time. The testimony of other women reveals that many thought they were paid poorly. However, in a way, it becomes relative, relative to the options available and to the other activities women pursued to support themselves and their families. Wages had to be placed in the context of overall subsistence activities. Cannery work allowed them to pursue other, non-monetary means of earning their livelihood. However, cannery work was essential because some form of wage income had become necessary to supplement other subsistence activities. When that source of money income disappeared, other income sources were also gone and the subsistence economy was no longer sufficient to support the population. State social assistance filled the gap.

Brenda Assu recalled how people used to congregate at Cassiar cannery, located on the Skeena River in the north, from Kitwanga, Kitsegeucla, Hazelton, and probably Kispiox (Skogan 1983, 37). Cassiar was the only operating cannery left in the early 1980s, its future in doubt since it had gone into receivership. Sunnyside, on the Skeena, was built in 1916 and closed in 1969. Mabel Ridley recalled how people came to work from up river, from Kitkatla, Port Simpson, Hartley Bay, and Kitimat. Most of them worked in the cannery all their lives. In addition to being a cannery worker, she was also a midwife. "I learned these things [midwifery] from my mother and my sister. At Sunnyside, when they first call me, I stay up all night maybe til the hard labour starts and the baby is born sometimes at five or six in the morning. Then I go to my house and tend to my own family and go to work. I was floor lady at Sunnyside for twenty years" (38).

Elizabeth Spalding began to work in 1916, when she was eight years old. Her mother was afraid to walk from the reserve to Port Essington (on the Skeena) by herself, so she took her daughter. She took a box for Elizabeth to stand on, and the women showed her how to do the work. She's been in the industry almost sixty years. She recalled Japanese women working there, with their babies strapped to their backs. "We don't know what rest is when we're young. When there's lots of fish we just quit for two hours to sleep, then work again. After a while we were paid around fifty dollars a ticket, but now they make real lots of money. Cheap in my time" (Skogan 1983, 63–4).

Leona Sparrow recorded the life histories of her paternal grandparents, both of whom were Salish (Fraser River area). Rose Sparrow also remembered Japanese women working in the canneries with their small babies strapped to their backs, while older children were kept in boxes in the cannery, where the women could keep an eye on them. Aboriginal women generally had either an older daughter or a grandmother to care for young children. Rose's twelve-year-old daughter looked after the children, one of whom was born at a Skeena River cannery. Rose left work at eight and the baby was born four hours later. She was allowed to return home to breast feed the baby because their home was close to the plant (Sparrow 1976, 93–4).

In June 1982, I travelled to Prince Rupert where a number of the researchers on the Fish and Ships project were conducting a sample survey and less structured interviews with fishers, shoreworkers, fishing company managers, and others connected to the industry. The testimony of aboriginal women as gathered by Sparrow, Skogan, and

Steltzer and Kerr was confirmed in our own interviews with aboriginal women cannery workers. We spoke with a number of women who identified themselves as Tsimshian who had spent most of their working lives in the various plants. They told us how they began to work at very early ages and had spent all of their working lives in the industry.

For example, one woman only had seven years of schooling before she started canning crabs in the 1969 season. Then in June, 1970, she began to work for B.C. Packers in its Port Edward plant and had been there ever since. She was raised in Port Edward and her mother had also worked in the cannery while her father was a night watchman there. She told us about the numbers of women working (700 to 800), about her various jobs (like washing fish, canning clams), and about the seasons: herring from April to May, salmon from June to October, and the rest of the time collecting unemployment insurance, waiting for the cycle to begin anew. Except that Port Edward was being closed. In June of 1982 she began work at Oceanside, where B.C. Packers had amalgamated its northern cannery and other fish processing operations. But she reported to us that she had obtained her job there thanks to her older sister. Including UIC and family allowance, she estimated she earned $18,000 a year gross (she made $13,630 from her cannery work in 1981). It is also interesting that she referred to her job as "Iron Chink," referring to her work on the butchering machines. She also provided us with the names of two Chinese bosses from Port Edward, thus also confirming the prevalence of the contract system and the legacy of the "China gangs." She noted that over 100 Tsimshian women were working with her at Oceanside. Native women came from Kincolith, Ayansh, Terrace, Hazelton, and other places. She confirmed for us what others had also told us; that is, that in the large plants like Oceanside, aboriginal women worked together according to the villages from which they came. Thus, in the face of northern cannery closures, along rivers like the Skeena, women still reproduced their communities in the ways they chose to work together on the lines in the huge Oceanside plant. Another Tsimshian woman, from Kincolith, told us how she had begun work in 1974, when she was still in school. Her mother and sister also worked with her in the Port Edward plant. However, when B.C. Packers bought Oceanside and amalgamated canning operations there from Port Edward, her seniority was too low and so she went to work in the Prince Rupert Co-op. Again, she noted how women from various ethnic groups worked together. Portuguese women were mostly filleters while the East Indian women worked in the canneries in their own groups. Another important point to note from the testimony of the aboriginal women, confirmed by others like foreladies and plant managers and evident in the seniority lists themselves, was the seniority of these women. There is a myth that in seasonal work like fish canning, there is a high turnover of labour. The fact that a number of aboriginal women have spent over forty years in the industry, alongside the history of intergenerational and family patterns of cannery work, shows that these women have devoted their lives to an industry that has often taken their devotion and their work for granted. There appears to have been little consideration on the part of large companies like B.C. Packers on how plant closures and amalgamations would affect the lives of women in numerous villages that had come to rely on canneries for their livelihood.

This section has discussed the role of aboriginal women in the canneries although it is obvious that the discussion necessarily entails the employment of Chinese men and the Chinese contract system that incorporated aboriginal cannery workers in individual plants. Until unionization in 1945 the contract system was the most widely employed method of hiring the inside labour in the plants. Despite the formation of the United Fishermen and Allied Workers Union in 1945 and the formal abolishment of the contract system as contravening post-Second World War labour laws, the system survived.

HISTORICAL BACKGROUND TO INITIAL CHINESE MIGRATION

"The original Chinese migrants were part of a great overseas diaspora that resulted from the convergence of two major historical forces: a rural crisis in China and Western imperialism" (Tan and Roy 1985, 3). The first Chinese migrants went to British Columbia by way of California and San Francisco, where the first Chinatown in North America was established. Many of these men, and the migrants were overwhelmingly male, entered mining, following the gold rush from California in 1848 to the Fraser River a decade later. Tan and Roy (1985, 7) note that most of these men "engaged in placer mining, an activity requiring limited capital investment, and confined themselves to diggings abandoned or sold by white men as being no longer profitable." However, in "colonial British Columbia" the Chinese enjoyed "full legal equality" (7). That is, the first provincial statesmen worked to actively *deprive* them of the franchise. They were supported in their efforts by federal politicians and various organized groups, including white male unionized workers. By 1876, at all three levels of government (municipal, provincial, and federal), "the Chinese were legally excluded from participation in the electoral process even though some of them were British subjects by virtue of birth in Hong Kong or by naturalization" (Roy 1989, 46).

> Chinese migrants came predominantly from the southern coastal province of Guangdong and the neighbouring coastal province of Fujian. Both areas were in the van of Western efforts to break down the cultural and commercial barriers surrounding China from the early nineteenth century. Canton, the Guangdong provincial capital, was the point of contact with European commerce before the disastrous First Opium War in 1839 ... It was through the original treaty ports in Guangdong and Fujian and the British and Portuguese possessions in Hong Kong and Macau that Chinese went abroad ... The overwhelming majority of Chinese migrants to Canada have come from a small area of eight contiguous counties in the heart of the Canton delta. (Wickberg 1982, 7)

The first Europeans known to have established trade with China were the Portuguese. They went to China in the sixteenth century and developed a triangular network of exchange involving Europe, China, and South America. At that time the Chinese state was insular and generally hostile to foreign trade. Government officials tried to confine trade to coastal ports. Canton emerged as an important port, and this had an impact on the surrounding countryside in Guangdong province. The peasantry began to specialize in the production of sugar, textiles, porcelain, and metal wares, which were exchanged for South American tobacco, sweet potatoes, and peanuts. The new foods were adopted into the peasants' diet, and prosperity from foreign trade led to population increase (Wolf 1982, 256). Tan and Roy (1985, 3) estimate that the population of China more than doubled during the eighteenth century, with population pressure especially acute in the south, particularly in Guangdong.

A change in dynasty in 1644 resulted in a tightening of imperial control over foreign trade. The Manchus were alien rulers who had a difficult time legitimating their power. While they despised foreign traders, they needed the gold and silver that were coming into China. In 1757 they decreed that Canton was to be the only port open to foreign trade. A group of merchants, the Cohong, controlled this trade, and they acquired a monopoly position when the emperor assigned them the regulation of the activities of foreign traders in China (Chan 1983, 22–3). In 1644 the Dutch introduced tea to England, and the subsequent European demand allowed China a favourable balance of trade. Since there was no reciprocal demand from the Chinese in the way of products, the British were forced to pay with silver, taken from the mines of Mexico. Wolf (1982, 255) estimates that, between 1719 and 1813, one fifth of all the silver extracted from Mexico and about the same amount taken from European silver stocks ended in China, which became the "tomb of American treasure." In searching for articles the Chinese might desire, European traders found that sea otter skins fetched high prices in China. In this way, the First Nations of North America were drawn into a triangular trading network controlled by the Europeans.

In 1776 the American revolution cut England off from its Mexican silver supplies. However, three years previously the East India Company had established a monopoly in the sale of opium, which the British promptly introduced to China. Opium addiction burgeoned, and the balance of payments shifted to the detriment of China. "The outflow of silver from China soon affected the country at large. The government set tax quotas in silver; the peasants paid in copper cash. As silver grew scarce and rose in price, ever larger amounts of copper were required to meet taxes. Opium thus did more than undermine the health of Chinese addicts; it began to subvert the social order in the countryside" (Wolf 1982, 258).

The peasantry in southern China was especially hard hit, since it had profited from foreign trade, substituting imported food products with articles destined for export. Along with opium, the British also sought to establish markets for their manufactured products, including textiles (Bolaria and Li 1985, 83). These flooded into China at prices Cantonese textiles producers were unable to match. While the Portuguese had developed mercantilist trade networks involving the exchange of goods among countries, the British sought to create markets for their manufactured goods and sought overseas centres of supply in resources necessary for capitalist production at home. One can draw a loose comparison here between the relative advantages enjoyed by the peasantry during the mercantilist phase and North American First Nations during the mercantilist period of the fur trade. The big difference is that the First Nations were in the process of being colonized during the fur trade, while the Manchus managed to keep foreigners out of China during the mercantilist period of European expansion.

"With the re-establishment of imperial control, foreign traders were again welcomed at Chinese ports." Over the course of the century from 1685 to 1760, the English East India Company increased trade with Canton, in preference to other ports open to external trade. "When the emperor again restricted foreign trade in 1760, Canton remained the only port open to foreign trade" (Wolf 1982, 254). However, unlike the earlier period of trade which preceded it, the introduction of "foreign capitalism" in China in the course of the nineteenth century had negative consequences for the peasant economy situated in the Canton region and supplying it with trade items (Bolaria and Li 1985, 83). "On the one hand, it [foreign capitalism] destroyed the local handicraft industries in villages and cities, and replaced them with a commodity market; while on the other hand, it drove many peasants and handicraftsmen to bankruptcy, and produced a large pool of surplus labour (Mao 1967). It was largely the desperation of starving peasants and frustrated intellectuals, for example, that led to the outburst of the Taiping Rebellion (1850–64)" (Bolaria and Li 1985, 83). Peasant families could no longer engage in cottage industries that produced internationally valued commodities. Instead, they were forced to produce agricultural surpluses convertible into copper and silver in order to pay the ever-heavier tax burdens imposed upon them by the upper classes. They received nothing in return. Worse, earlier population growth acted as a severe barrier to the peasantry's ability to survive on the land. Younger sons who had little hope of inheriting land sought wage employment in Canton as a way to help their families. But the industries that could hire them were also facing severe economic depression. The only solution for many of these surplus labourers was emigration. However, between 1672 and 1858 emigration was prohibited. Foreign demand for this surplus labour supply and the foreign military presence forced legislative changes and partially determined the pattern of emigration.

The Chinese state's resistance to England's attempt to colonize the country and to introduce opium as a commodity led to the Opium War, begun in 1838 when the British navy sank three Chinese gunboats. In 1842 a defeated China was forced to sign the Treaty of Nanjing. "The impact of Western military and economic invasion was to disrupt the feudal autonomous system in China and replace it with a neocolonial structure" (Li 1979, 322). Canton lost its privileged position now that other Chinese ports were opened to European traders. Foreign domination was matched by internal political disintegration as the emperor lost his ability to control the warlords. Warlords jockeying for position, fighting one another to overthrow the state, forced the peasants to finance their wars with higher and higher taxes (Wickberg 1982, 9). The result was a series of peasant revolts, the most famous of which, the Taiping Rebellion, occurred in 1850. The Qing dynasty, with foreign aid, eventually crushed it in 1864, with an estimated loss of 20 million lives (Chan 1983, 35).

China was a patriarchal as well as a feudal society. Sons contributed to family income and carried on the lineage. Nineteenth-century China was governed by Confucian ideology which subordinated women to men from birth to death. Women could not own property, hold political office, or play an important role in religious worship. Thus, in times of economic distress, while men sought to continue the family lineage by seeking wage employment and by even emigrating overseas, women were killed in infancy, married at an early age, or sold into slavery (Yung 1987; McCunn 1988). For example, one Chinese proverb stated that "eighteen gifted daughters are not equal to one lame son." "During the mid-nineteenth century, when Chinese immigrants began coming to America in large numbers, Chinese women were destined to perform household duties, marry and bear sons, and serve their husbands and in-laws. Dutiful wives were left at home to tend to the family while their husbands went overseas in search of fortune" (Yung 1987, 11).

It is paradoxical that Chinese male emigrants of the latter part of the nineteenth century – who were a product on the one hand of class relations that created a surplus labour pool of Chinese peasants due in no small part to European imperialism and on the other hand were conforming to Confucian ideology that severely restricted the movement of women out of China – were in their turn subjected in North America to both class exploitation as cheap wage labourers and to a racism connected to European ideology that held the white man superior to all women and to all non-white men. As a result, many of these emigrants came to perform labour designated as "women's work" in European societies: in laundries, restaurants, as domestics and on assembly lines in the factories (for example, the manufacture of boots and shoes, cigars, and the wholesale manufacture of clothing) (Canada 1902, "Royal Commission on Chinese and Japanese Immigration" [hereafter referred to as RC 1902]). Their employers claimed they hired these men because white women were in short supply and the economy was booming, creating a demand for the labour power of those whose labour had in previous times gone unvalued – women.

As demonstrated in chapters 1 and 2, while women's work had previously been unvalued, it now became part of the labour necessary for the development of the capitalist mode of production in North America. Women's work, that part now incorporated within capitalist production, became public and visible and thus acquired value in that now it had to acquire at least some monetary value, although it continued to be judged inferior to the labour provided by white men. However, Chinese labourers (and Japanese fishers who entered the province in growing numbers from the 1890s) were steeped in their own patriarchal culture which also assigned value to being male and thus to activity designated as men's labour. There was nothing in their "conscience collective" that prevented them from entering and successfully competing with white men (on the contrary, in China and Japan there was a history of disdain and distrust of the white foreigners). However, their material conditions enabled them to compete by dragging down the price of wages and thus disrupted the European patriarchal valuation of "men's" and "women's" work. Both Chinese and Japanese men entered both types of labour. Especially instructive in this regard are the chapters on agriculture, market gardening, the coal mining industry, placer and lode mining, the lumber industry and shingle business, and merchant tailors contained in the 1902 Royal Commission report. Even more threatening to white capitalists, the 1902 Royal Commission cites evidence of Chinese and Japanese men becoming successful merchants and capitalists, thus ignoring class and race barriers.

While the Chinese and Japanese were competing with white capitalists, merchants, and the working class, they were also judged in testimony submitted to the two Royal Commissions (published in 1885 and 1902) as well as in the British Columbia Fishery Commission (whose findings were published in 1893) as preventing the immigration of white women and children of the working classes into the province, thus preventing the development of a "white man's province." The term "cheap wage labour" was used in the testimony to mean exactly that which has been defined here in earlier chapters: labour that white men did not want to do because it was disagreeable and because it was priced at a level that did not allow white men to raise families in the province. Chinese and Japanese men could fill these positions because they had immigrated as "solitary males" and were not perceived to be committed to remaining in the province. Many white men testified to the commissions that these positions would be ideally filled by white women and white children, thus allowing settlement of British Columbia by a white proletariat that would in turn help industrialize the province with an indigenous (white and aboriginal) labour force. Point four of the resumé in the report of the Royal Commission on Chinese Immigration (Canada 1885, cxxx [hereafter referred to as RC 1885]), for example, states: "That the Chinese do largely engross domestic service ... white girls cannot be induced to go into the country, removed from their church and accustomed companionship, to work as domestics, and that a sufficient supply can not be had even for cities and towns; that the fact that the Chinese compete with female servants is, nevertheless, one well worthy the attention of Government."

British Columbia historians (Ward 1978 and Roy 1976 and 1989, to name two of the most prominent) trying to explain why British Columbia has witnessed such virulent racism towards its Chinese and Japanese populations have tended to focus upon either psychological

(Ward) or economic (Roy) explanations. My argument, based on the fishing industry and on the creation of cheap wage labour, suggests a far more intricate and interlocking set of forces that include the psychological and the economic but that also go beyond these. At issue during this period was a power struggle over which groups would direct the capitalist development of the provincial economy. Thus it was not only the white male proletariat that felt itself threatened. White capitalists could easily be displaced by capitalists from China and Japan who had recourse to considerable investment from profits taken from mercantilist activities overseas. While capitalists, like salmon canners, did everything in their power to preclude their own displacement, at the same time they desired to have on hand cheap wage labour forces from these same countries. It was convenient, then, to pit and divide the working class against itself by emphasizing racial distinctions, although this characterization lends itself too easily to a deterministic analysis. European men already saw such differences as important. In particular, to assign Chinese men to "women's work" was also to degrade Chinese men as men, thus rendering the boundaries between sexism and racism permeable, reinforcing the latter by recourse to the former.

Chinese and Japanese immigrants did not roam the Pacific Northwest as solitary individuals in search of waged employment. Their migration was part of a well-organized pattern controlled by merchant companies who entered the immigration process and acted as mediators on behalf of the home governments, foreign governments, the migrants' families and communities, and overseas employers.

THE CHINESE CONTRACT SYSTEM

Johnson and Wickberg (Wickberg 1982, 5) identify two patterns in the first period of Chinese emigration: coolie broker and chain migration. The Chinese contract system was an extension of the first. Since peasants could not afford to pay for the voyage overseas, they went into debt with a broker who paid their passage fare. The immigrant thus became indebted from the start to the broker and he could not seek employment on his own until the debt was repaid. "Passage loans were often made by Chinese clan associations established in California, known as the 'six companies,' and later by capitalist employers such as the railroads who recruited for labor in China" (Boswell 1986, 358).

Mention has been made of the Cohong merchants who controlled foreign trade through the port of Canton until the outbreak of the Opium War. In feudal China and Japan the merchant class was considered inferior to the scholar-gentry, to artisans, and even to the peasantry. But the status of this class rose with industrial development. Merchants also took on the role of middlemen in economic development at home and nascent capitalism overseas. When they lost their monopoly over the distribution of trade goods leaving Canton, Cantonese merchants turned to supplying labourers for the plantations of Hawaii, Cuba, the West Indies, and South America. Since the trade was illegal at that time, it involved the forceful abduction of Chinese peasants, and in this sense the coolie trade was similar to the slavery it replaced. But it only lasted for a short time.

The slave trade was abolished in the United States in 1808, although slavery itself did not end until 1865, following the Civil War (Wolf 1982, 315). The coolie system flourished briefly when the slave trade dried up, but it ended when the Chinese state legalized emigration. To be effective, however, the countries importing labourers also had to stop abuses. In the 1870s Spain and Portugal agreed to regulate immigration from China. The United States chose a different route. Just as England had wanted to open China as a market for its goods and a supplier of raw materials for its industries, so the United States wanted to open China as a source of labour supply. It achieved its end in 1868 with the signing of the Burlingame Treaty. Henceforth, Chinese emigrants were to be allowed to migrate freely. A most-favoured-nation clause in the treaty covered Britain and its former colonies. These series of events ended the coolie system in that peasants were no longer forcibly and illegally transported from China. But the system of contracting cheap labour in China to meet American demand continued: "The contract labor system demanded more accountability than the coolie traffic (which saw hundreds of indentured workers die of scurvy, malnutrition or flogging en route to the New World), if for no reason other than that the investment outlay was greater. The worker's value was also considerably higher because he had signed on by choice and was travelling as a free emigrant. After 1870, the contract type of labor was the only legal way a Chinese laborer could work in British colonies and former colonies" (Chan 1983, 45).

Boswell (1986, 358) argues that the "six companies" acted as middlemen in the interests of the capitalist class, providing "low-paid labor." These companies acted as an unofficial arm of the state by enforcing contracts and "policing Chinese behavior." The peasants were forced to work for these contractors for an agreed period in order to clear their debts. Most of them planned to return to China rich men, and left their families behind, planning to ease poverty at home with remittances from overseas pay cheques. But a whole pattern of indebtedness developed between the labourer and the merchant class resident in China, marked by the establishment of a class of merchants and agents abroad. The peasants left by the port of Canton and had to pass through Hong Kong. Many of the consuls there were involved in the trade in labour, as is evident from the evidence collected by a U.S. inquiry into Chinese immigration in 1876, and appended to the 1885 Royal Commission (RC 1885, 179–360).

Thomas H. King, a San Francisco merchant, lived in China for ten years, was acquainted with the activities of the consul's office in Hong Kong, and assisted in the placement of Chinese emigrants on ships leaving for North America. His testimony provides some clues as to how the system worked. The laws of the U.S. government apparently allowed the consul to exact fees from each labourer. Nearly all of them, except a handful of boys and Chinese men returning to California, appeared to be under contract. The contract bound them for three to five years, but few appeared to know the conditions of their employment overseas. Part of the contract stipulated that they would be cared for if sick, and would either be sent back to China at the expiration of their contract or their bodies would be shipped back if they died. Money advanced to cover the cost of the voyage had to be repaid with interest running as high as 5 percent per month. The contractors also arranged for wages to be remitted to families in China, and this created the opportunity for further abuse. King claimed all sailing vessels to China had conditions in their charters stipulating that the only labourers to be taken were those supplied by the companies. Although force was not exerted, many peasants went unwillingly, to clear debts, to support family and friends at home, or to better their own situation. The contractors at the San Francisco end were originally agents of Chinese firms in Hong Kong, who in turn were connected to firms in Canton. The companies in San Francisco were known as the "Six Companies," representing the six districts around Canton from which labourers were recruited (RC 1885, 188–95).

Yee has written two articles based on his research of the records of Sam Kee, one of the wealthiest merchant companies in Vancouver's Chinatown in the first decade of the twentieth century. He includes information on labour contracting in the salmon canning industry. The founder of the company, Chang Toy, arrived in Victoria in 1874, his passage paid by a fish canner for whom he was to work for one season in repayment. Because the boat arrived late, only one month's work was left, and Chang then went to work in a sawmill in New Westminster, gradually moving into other lines of work, including "labour contracting in the timber, fishing and sugar industries." Sam Kee provided contract labour to, among others, John Wallace, J.H. Todd, and the Windsor and Imperial canneries (Yee 1986, 73, 86).

> In the salmon industry, a cannery operator and a Chinese labour contractor drew up an agreement whereby the contractor agreed to supply men at a specified price to can a certain quota of fish. The canner provided a cash advance, which the contractor used to entice men to take the job. Crews were then sent to the canneries where they worked and were boarded by the canners. At season's end the contractor received the final payment and paid his crew their wages less amounts deducted for the advance, room and board.
>
> However, if the salmon run for that season was low, or if any other reason prevented the workers from filling the quota of fish stated in the contract, the canner could refuse to pay the balance of the contract. The contractor too could pass his loss onto his workers by inflating the costs of provisions and food he had supplied. Thus workers could emerge from a cannery having been fed and housed but denied their expected wages. In short, the contract system shifted the risks of an unstable industry onto the contracted workers and their contractors. (Yee 1986, 85–6)

The companies operated within two modes of production. They offered services required by Chinese peasants that were not available to them overseas. In the mid-twentieth century many of these services would be taken on by the welfare state, but in an inhospitable and hostile environment Chinese labourers depended on the benevolent associations and social services provided at least in part by the merchant class. But the primary occupation of these companies overseas was to make profit by supplying the needs of a developing capitalist economy. Thus, while they provided services essential to Chinese immigrants, their profit-making motives and their hold over these men, most of whom could not speak English, provided the opportunity for abuse and for exploitation beyond their role as suppliers of cheap labour power. We are not only dealing with class exploitation here but also with a whole hierarchy of status relations tied to Confucian ideology and the importance of family and lineage in this system.

Chan (1983) makes a convincing case to the effect that Confucianism and the class structure in China resulted in a hierarchy of authority modified by the merchant class to suit North American conditions. The Confucian system of values placed central importance on the family. And the family, including ancestors, represented "a microcosm of society and the state." Both were ordered in a rigid structure of authority: "The grandfather, typically the oldest living male, governed a patriarchal hierarchy that stressed the subordination of the son to the father, wife to the husband, younger brother to the older brother and in civil society, the subject to the ruler ... The authority of the oldest male adult, whose central purpose was to enrich the family and enhance its prosperity, gave him almost total control over the destiny of his immediate household and families related by kinship as well as those with clan ties" (100–1).

Capitalist employers in the United States, Australia, New Zealand and Canada used Chinese labour because it was cheaper than most other categories of labour. The Chinese peasant subsisted close to the margins of starvation in the depressed conditions of southern China, and he travelled alone. Thus his needs were marginal, well below the standards achieved by working class men coming from western Europe.

However, the production and reproduction of both groups of labourers continued to be met outside the capitalist mode of production. In the nineteenth century, many immigrants from western Europe also travelled alone. They expected to find work in North America at wages higher than those they could earn at home. Since they did not have families to support them, they needed higher wages to buy the commodities necessary to survive, and at the same time expected to develop a higher standard of living than they could hope to find at home. Once that was attained, they planned to send for their families or find wives and establish these in some comfort, at least with prospects of increased wealth in the near future.

> To this day British Columbia retains a distinctive identity within Canada. To the creation of differences between the province and the rest of the country, many factors have contributed, including climate, geographic isolation and a semi-independent economy. Another factor, both cause and effect, is the existence of a separate ethos ... If working-class immigrants helped create consciousness of class in British Columbia from the bottom up, their middle and upper-class compatriots have done no less from the top down. (Barman 1986, 45)

But the production and reproduction of the labour power of Chinese labourers continued apace in a depressed Chinese economy, and thus the subsistence needs of Chinese labourers could be met at wages impossible to enable European men to subsist and reproduce. Labour contractors maintained the connection between labourers and their families left behind in China. This enabled the contractors to offer labour power at a rate far below the standard needed to maintain white male labourers. The reaction of white male labourers to this threat was rooted in patriarchal and colonial ideology. Rather than organizing with fellow (non-white) proletarians, they refused to see any similarity in their joint conditions. Their response was to organize collectively *against* this group of workers. It helps explain the extreme racism that characterized British Columbia at the turn of the twentieth century (Roy 1989). And this worked in the interests of employers. Because the Chinese were categorized as different on the basis of race, they could continue to be exploited as cheap labour and excluded from efforts to unionize white male workers. Nevertheless, Chinese labourers developed their own labour unions, and organized to protest the abuses they suffered from the contractors. For example, in 1889 a delegation of Fraser River salmon canners, reacting to a shortage of labour and a strike the previous season by Chinese labourers demanding higher wages, went to Ottawa to request removal of the head tax (Roy 1989, 67).

> Contrary to popular perceptions, the Chinese were not a docile work force. In 1881, railway construction crews rioted when a labour contractor tried to take a higher commission from their wages. Laundry workers also occasionally struck their Chinese employers for better wages and working conditions. Chinese also protested mistreatment by the white community ... The formation in 1916 of the Chinese Labour Association, a nonpartisan organization committed to workers' rights and "the abolition of the capitalist system," fused political radicalism and labour militancy. (Tan and Roy 1985, 10–11)

Chairman Gray notes: "By Provincial Legislation in British Columbia and the general hostility evinced towards them, the Chinese are practically prohibited from becoming attached to the country. They are made, so far as provincial legislation can go, perpetual aliens, and with the Indians are by positive means denied the political and municipal franchises attached to property and person, conceded to other British subjects, born or naturalized, when of sufficient age to exercise them" (RC 1885, xi). Senator McDonald of British Columbia testified: "If we had the number of women which they have in this part of the country [central Canada], they would do all that kind of light work, and then, of course, I would be in favor of doing away with Chinese labor altogether" (xxix). A few pages further on in the report evidence is presented that Chinese women were employed as prostitutes.

That most immigrants to the Pacific Northwest were men raised in patriarchal cultures encouraged the development of prostitution. Thus the position of women (aboriginal, Chinese, and European) was especially degraded. While the forcible abduction of male Chinese peasants ended with the coolie system, women continued to be sold into slavery to work as prostitutes. "Chinese women traditionally did not leave their home village and only about three per cent of the Chinese immigrants were women. The 'six companies,' which kept records of their clan members, estimated that 80–90 per cent of the approximately 6,000 Chinese women in California in 1876 were prostitutes, many of them were forcibly imported" (Boswell 1986, 67). While male peasants began to be treated as individuals, women continued to be bought and sold as commodities, fetching prices ranging from $500 to $2500 for the San Francisco prostitution trade:

> The "cribs," each of which held up to six women, were slatted crates, often located out of doors, measuring approximately 12 feet by 14 feet with a curtain, pallet, wash basin, mirror and usually two chairs. A woman forced into crib prostitution would work for six to eight years; at the end of her usefulness, when she was ravaged by disease, physical abuse or starvation, she was allowed to escape to the Salvation Army, the hospital or the gutter. Typically, she would be dead within six months. (Chan 1983, 81)

Prostitution was also handled through Chinese merchant companies, and there are some examples of women acting as merchants in this trade. They appear to have been connected to prostitution in China, advancing money to prostitutes there who then came to North America under contracts similar to those of male labourers (RC 1885, 192; and Exhibit 17 submitted to RC 1902, 39, a certificate of sale of a girl, Woon Ho, for $302). Many of those who were forcibly abducted were young girls of not more than thirteen years of age, and boys. As mentioned, however, the Chinese were not the only ones engaging in this trade. Fur traders had introduced prostitution to North America and paid aboriginal women for their sexual services. Women were also brought from Europe and the eastern United States, but they appear to have exercised more choice than those Chinese women, girls, and boys who were forcibly abducted. In the male-dominated society of the Pacific Northwest, it is indicative of the transplantation of patriarchal consciousness that the largest group of non-aboriginal women was originally engaged in prostitution, and that many were sold into the trade at a time when slavery was officially abolished.

RACIST CATEGORIZATION OF CHINESE LABOUR POWER

White male workers, settlers, municipal, and provincial officials all protested against the employment of Chinese labourers. Commissions were established in the United States and Canada as politicians attempted first to limit, and then to abolish, immigration from China. The major support for continued immigration came from employers. Their justification for the need to employ cheap wage labour illustrates the interrelationship between the categorization of gender and of race in structuring cheap wage labour within capitalist class relations of exploitation, and the transformation of patriarchal consciousness to widen the categories of those who are to be excluded from participation in the public realm, however it comes to be defined.

The reason frequently put forward by capitalists for employing Chinese men was the absence of white women and children. Boswell (1986, 359) notes: "The Chinese were particularly welcomed in positions that the white miners (who were almost all male) considered female work, such as cooking, housekeeping, or laundry." And the commissioners appointed to head the Royal Commission of 1885 noted in summarizing the evidence presented to them: "It is something strange to hear the strong broad-shouldered superior race, superior physically and mentally, sprung from the highest types of the old and the new world, expressing a fear of competition, with a small, inferior, and comparatively speaking, feminine race" (RC 1885, lxix). The commissioners went on to compare the Chinese to a "labor-saving machine" whose only difference from their mechanical counterparts was that they consumed the "productions and manufactures of the country, contributing to its revenue and trade, and at the same time expanding and developing its resources" (lxx).

Colonel F.A. Bee, consul for the Chinese government in San Francisco for almost six years, explained that the outbreak of the U.S. Civil War cut the Pacific states from their sources of supply just as they were embarking on capitalism. While raw material had previously been sent east for processing, now nascent industries were established. In the eastern United States and in central Canada, women were employed in these manufacturing enterprises; for example, in making boots and shoes, clothing, underwear, cigars, and matches and other light industries (Palmer 1983). Because of the scarcity of white labour in California, employers hired Chinese men in their place (RC 1885, 16–17). But these employers found that Chinese labourers were less easily subordinated than white women. While women had come to know their inferior position through centuries of patriarchal conditioning, Chinese men had until recently profited from a similar ideology to assume their own superiority in Chinese society. Thus several politicians complained that the Chinese proved to be "too good" at their work. Sir Matthew Begbie, chief justice of British Columbia, summarized the dilemma facing white employers: "The Chinaman is in every respect the reverse of an European, except that he is a man ... Yet they as evidently despise all our attainments and ways; and, what is most annoying, they come here and beat us on our own ground in supplying our own wants. They are inferior, too, in weight and size of muscle, and yet they work more steadily and with better success on the average than white men" (72).

In the development of capitalist class exploitation, it was the Europeans who initially had the upper hand, and it was their attitudes that shaped the power structures of North American society. But the fact that it was not the biological destiny of the white man to rule "inferior" races is proved in the fear white capitalists expressed that the Chinese could beat them at their own game. For example, the president of the Immigration Association of California, A.R. Briggs, noted that several industries were controlled by capitalists who were Chinese, and he gave as an example the manufacturing of cigars. Originally employed as cheap labour, the Chinese had learned the manufacturing processes. A few managed to accumulate capital and invested in their own factories, taking advantage of their connections with the Chinese merchant community at home and abroad. The Six Companies had the necessary assets and were in a position to finance these entrepreneurs. Briggs claimed that the Chinese dominated cigar manufacturing in California. "That is to say, cigars are made almost wholly by Chinese workmen, and many of the factories are in the hands of Chinese, and owned by them. They do the same thing in tin-ware, boots and shoes, and clothing" (RC 1885, 7).

Politicians and settlers promoting the immigration of women and children from Europe complained that it was difficult to induce white women and children to come when the jobs they could take were filled with Chinese workers. Even domestic labour, the epitome of "women's work," was dominated by the Chinese. In the process, however, the Chinese proved unsatisfactory because they too could become capitalists on their own account, unlike women of any group. The male white population resident in British Columbia perceived this as a threat to its own basis of power. The eventually adopted solution was to substitute machines for Chinese labourers and fill new positions with female labour power, enabling politicians to ban further Chinese immigration altogether.

THE CHINESE CONTRACT SYSTEM IN THE SALMON CANNING INDUSTRY

As noted in chapter 3, George W. Hume was the first salmon canner to employ Chinese labour – in 1872 at Eagle Cliff near Portland, Oregon. Stacey (1982, 2) dates the beginning of salmon canning on the Fraser River at 1867, when James Symes experimented with canning in his saltery. The only piece of machinery used was a large iron kettle to boil the tin cans. Aboriginal fishers had been supplying salmon to the Hudson's Bay Company forts for many decades and continued to supply salmon canners.

As demonstrated above, the canners could not rely on a totally aboriginal labour force for the inside work in the canneries because they were not assured that the labourers would remain at the plants for the entire season. There were frequent periods when there was no fish to process, and canners would not pay labourers unless they were working. Thus, if jobs or opportunities became available elsewhere, aboriginal labourers would leave the canneries. But once the runs started up again, the canners needed crews to process the catches immediately. Salmon canneries were often situated in fairly remote areas with poor communication. Thus the ideal labour force for salmon canners was one that would remain on site but that could be laid off during slack periods. The Honourable Justice Crease of the British Columbia Supreme Court remarked: "The Indians could not be depended upon at first on account of their numbers, which in those days were threatening, nor afterwards on account of their restless, nomadic propensities, which prevented them from settling down to any permanent, industrious avocations" (RC 1885, 142).

Aboriginal women and children continued to be employed, assigned to work that was paid the lowest wages. But the central tasks in the newly evolved division of labour were assigned to Chinese crews who were committed by the nature of their contracts to remain on site for the entire season. Cut off from their friends and families in China, and indebted to the contractors, these men had little alternative but to work and work hard. Originally hired because they were a reliable and cheap labour force, they eventually became a highly skilled one. Since most were of peasant background, they did not arrive with these skills but developed them on the job. As the salmon canning industry grew from its initial experimental stage, certain functions became crucial to the overall operation, including butchering the fish, cooking it properly, making the tin cans to contain it, and testing the finished product for defects (defective cans could lead to botulism; botulism scares, occurring periodically throughout the history of this industry, could lose canners entire markets). Chinese crews became expert in all these tasks.

To provide an idea of the actual division of labour that took place in the canneries, I will give a brief sketch of the labour process, pieced together from a handwritten account by Alfred Carmichael and entitled "Account of a seasons work at a Salmon Cannery" (Windsor Cannery, Aberdeen, on the Skeena River, 1891) and from T. Ellis Ladner's autobiography in *Above the Sand Heads* (1979) and *Mosaic Fragments* (1980). While Carmichael's account focuses on the north, Ladner gives a first-hand account of pioneering life in the south in what became the Municipality of Delta and includes the Fraser River. Because of the costs involved in transporting workers by boat from the south, proportionately more aboriginal employees worked in the northern canneries.

The canning season on the Skeena River began around the middle of June. During the winter, the manager, based in Victoria, contracted with a Chinese firm to make the cans and process the salmon. When the river was clear of ice, around the beginning of April, the Chinese labourers were sent by steamer to Windsor.

Windsor Cannery had two main sections: a cleaning house with two double cleaning tables and a filling room containing three tables with benches at the ends, at which aboriginal women worked. At the first cleaning table, the head, tail, and guts of the fish were removed. The remains were shoved into a hole at the center of the table, from where they fell into the river. The fish was then sent to a washing tank. It was scraped, scrubbed, and the fins were removed. The slimer scraped loose scales and the inside of the fish with a knife, and then scrubbed it with a brush. After a second washing, a cutter placed the fish in a manually operated gang knife. He cut the fish into lengths. The sections were then lifted by hand to chopping blocks, and Chinese butchers split them into sizes for canning. The butchers slid the fish down an incline into a bucket. When the bucket was full, it was carried to the aboriginal women for canning. A little salt, deposited on a board, was placed in each can as it was filled. Women were paid by the number of tickets, one per tray of cans filled, they collected. The number of cans per tray varied with the size of the can.

The trays were weighed and cleaned. Aboriginal women weighed the cans, while cleaners, usually aboriginal children, wiped them with pieces of old netting (the ratio appeared to be six children to one or two women). Chinese men put on the lids and fastened a piece of tin under the hole in each lid. The tops were crimped to the bodies, and then placed on a conveyor belt to a machine where Chinese men soldered the tops. Once through the machine, they were poled down a shoot to the "bathroom." Chinese men placed the cans on coolers, on hand trucks (tables with wheels). They soldered the vent holes in the cans, and then the "bathroom crew" took over. This crew consisted of skilled Chinese who had made the cans earlier in the season. The cooler was hooked onto a crane and then run along to the first tester. Steam was turned on, the water boiled, and the coolers lowered into it. With the heat, the contents of the cans expanded. Air would escape from cans with any leaks, which were lifted and sent to the leak stoppers. The cooler was then raised and lowered into the first boiler (three boilers, accommodating sixty-eight cases, were filled). These boilers (or retorts) were filled, sealed off, and steam was turned on them for seventy-five minutes. After cooking, the steam was turned off, and the first bath emptied.

After cooking, the coolers were sent to the testers. These Chinese men pricked holes in the tops of the cans to let out the steam. A few drops of flux were applied to seal the cans, and they were then sent back into the boilers for a second cooking. Ladner (1979, 107) provides a vivid description of the Fraser River operation: "The coolers were stacked on low-wheeled, steel-framed cars running on steel shod tracks. The cars were run into large steel retorts, set in a row and serially numbered, with a turntable in front of each." The cars were then pulled out, placed on the turntable to remove the water, and pushed to the washers. The Windsor cooking operation appeared to be more manually intensive than those in most Fraser River canneries. Carmichael noted that, at Windsor, three aboriginal men washed 1300 cases for $90. Each cooler was lifted and dipped into a caustic soda bath that removed the grease adhering to the cans. The cans were cooled for one day. Chinese labourers then sounded each can with a small steel rod or large nail. They could tell by the sound which cans were defective or underweight. These were removed, repaired, or discarded.

When fishing ceased, the Chinese crew worked at casing the pack. Half the labourers made boxes, while the other half labelled and lacquered the cans. At Windsor, they were paid 37 cents per thousand cans. The cans were sounded once again for defects, and finally cased. As mentioned previously some of the canneries had sawmills attached to provide lumber needed for tasks like casing the packs.

There was some variance in the use of labour. For example, Ladner noted that Indian women were almost never hired as fillers, whereas on the Skeena they were. At Wellington Cannery, women were usually employed as slimers, to wash the fish, while wipers were Chinese, many of them "mere lads" (Ladner 1980, 56). This work was boring and often done by children. Assignment of tasks to specific groups depended on the relative sizes of each labour force. However, tasks were assigned by racial distinctions and there is no evidence, for example, that aboriginal women and Chinese men worked together on the same tasks in any one cannery (Muszynski 1987, 61–2).

The Chinese contract system served canners well in providing them with a labour force that would remain on site and available no matter what the hours, which ranged from no work to requiring crews twenty-four hours a day. And it was left to the contractor to pay his crews at the end of the season, relieving the salmon canner of the need to hire and pay individual employees. In the testimony to the royal commissions of 1885 and 1902, salmon canners and others affirmed over and over that without the Chinese the provincial salmon canning industry could not have developed.

To give an idea of the size of the Chinese labour force engaged in salmon canning, New Westminster (located on the Fraser River) had, in 1884, a permanent Chinese population of 300, which swelled to between 1200 and 1500 during the fishing season. Labourers came from Oregon, Washington state, California, and Victoria.

By 1900 an estimated 6000 Chinese were employed in provincial canneries. Henry Bell-Irving provided information for the English syndicate, Anglo-British Columbia Packing Company, on the labour forces employed in its canneries. In 1900, the six canneries on the Fraser employed from 700 to 1200 inside workers (these two figures demonstrate the enormous fluctuation in the number of labourers required at any one time). Of the 1200, 180 were white, 300 were Indian women and the rest were Chinese (see table 6).

> At two of the canneries on the Fraser River inside work is done by day labour. It is done by contract in the others. The Chinese contractors hire their help in their own way; we do not generally inquire how. Approximately their wages vary from $35 to $40 a month. They board themselves. Indian women are paid by piece work, as a rule, for filling cans. They earn from a dollar to a dollar and a quarter a day, sometimes more. They are hired by the Chinese contractor. Scarcely any Japanese and Indians are employed inside the canneries. The proportion of whites to Japanese and Indians employed in and about the canneries is about the same. We employ fewer Chinese on the Skeena, and there is a larger number of Indians and Indian women inside the canneries. We employ there about seventy-five Chinamen in each cannery, about fifteen white men and seventy-five Indians, male and female. (testimony of Henry O. Bell-Irving of Vancouver, RC 1902, 143)

The stereotyped image of the Chinese labourer as docile and passive was contradicted by the various collective efforts Chinese labourers made to improve their lot. As mentioned earlier, after the provincial government attempted to pass a labour tax on Chinese in Victoria, all Chinese servants employed in that city went on strike in 1878. The federal government disallowed the act. Briggs noted that in San Francisco, when the Chinese began to realize the value of their labour, they tended to organize "very much as the whites do in trade organisations, and strikes among those people are as frequent and as arbitrary as among the whites" (RC 1885, 8). After the turn of the century, head taxes and other state regulations cut back the number of Chinese

Table 6
Wages Paid to Chinese and White Men by Ewen & Co., 1897–1900

Chinese

Year	*Month*	*No. of Men*	*Average Time per Man per Month*	*Average Monthly Earnings per Man*	*Total Wages Paid for Month*
1897	April	45	16 days	$23.38	$1052.10
	May	45	13 days	18.93	851.85
	June	45	22.5 days	31.71	1426.75
	July	155	16 days	24.27	3761.85
	August	159	23.25 days	35.77	5687.43
	September	53	20 days	30.12	1596.36
1898	April	9	1.8	2.75	24.75
	May	38	24	33.06	1256.28
	June	38	24	34.12	1296.56
	July	116	6	8.67	1005.72
	August	116	13	19.70	2285.20
1899	May	4	6	7.14	28.56
	June	35	6	9.50	332.50
	July	122	7	10.96	1337.12
	August	146	14.5	22.00	3212.00
	September	44	16	25.40	1117.60
1900	April	12	2	3.33	39.96
	May	30	21.5	33.09	992.70
	June	30	9.5	16.26	487.80
	July	63	6	9.56	602.28
	August	85	16	22.69	1928.65

White Men

Year	*No. of Men*	*Average Time per Man for Season*	*Total Wages Paid for Season*	*Average per Man for Month of 26 Days*
1897	19	5.5 months	$8316.23	$79.58
1898	21	5	7950.51	75.71
1899	20	5	7720.95	77.21
1900	20	5	8091.71	80.91

Source: Report of Royal Commission on Chinese and Japanese Immigration, in *Sessional Paper* No. 54, 1902, 140–1.
Ewen's Note: "White men are paid from $40 to $100 per month and board – above figures include board at $12 per month. Chinese are paid for actual time worked only, and in all cases board themselves. Their wages vary from $35 to $75 per month."

allowed to enter Canada. The proliferation of canneries led to a demand for and scarcity of Chinese labourers, who then began to insist on higher wages and organized to protest abuse from their contractors.

A number of canners provided more information on the workings of the contract system to a fishery commission held in 1892. F.L. Lord, owner of Wadham's cannery, noted that Chinese contractors also hired aboriginal women. He had visited the canneries on the Columbia River and claimed he had never seen a white man employed in the type of work done by the Chinese since "a white man would starve to death" (BCFC 1892, 179). R.P. Rithet, who identified himself as involved in the "local" (i.e., no outside capital) syndicate, testified that the salmon canning business could not have been carried on without Chinese labour (269–70). It was impossible to obtain white labour for such a short time, and the work itself did not require "able bodied men" – it could be done by Chinese men, by women, and by boys (275).

The Chinese contract system persisted until around 1949, at which time union agreements gradually began to replace it (remnants of it, however, continued well into the 1980s). Labour legislation introduced during the Second World War required employers to place workers on payrolls and to record the wages paid to each individual. This in effect rendered the contract system obsolete. In the 1980s a former contractor working for B.C. Packers in Prince Rupert described how he recruited workers. "My job was a labour contractor – to supply all the labour and be in charge of cannery production. Got paid by the case. No, you didn't have to be Chinese to work for the contractor. My father hired Indians too" (Skogan 1983, 76).

The canners provided Chinese crews with accommodation in the form of bunkhouses. Conditions in many of these were abysmal. Men were often packed into bunks with little room between them and the facilities (toilets were often nothing more than holes in the ground) were generally inadequate for the number of men using them. When Japanese fishers were initially hired, they too were put up in their own bunkhouses. White male employees had the best accommodation. They and their families often lived in cottages set aside for them on the grounds. Finally, aboriginal labourers and fishers camped in their own groups, either in tents or in accommodation provided by canners. Several families would often be crowded into a two- or three-bedroom cottage. At each cannery the living arrangements mirrored the racial segregation inside the plants and on the fishing grounds, with the quarters of each group spatially separated from those of the other groups.

At the turn of the century, a large proportion of the Chinese and Japanese populations residing in British Columbia were concentrated in the fishing industry. There were 74 canneries in B.C., 49 of them on the Fraser. The Chinese Board of Trade of Victoria estimated that of a total of 3263 Chinese labourers in that city, 886 were cannery men, the largest category. The second highest number, 638, was employed in miscellaneous jobs, while an additional 530 labourers were domestic cooks and servants employed by whites (RC 1902, 12). The fishing industry employed an estimated 20,000 people, 10,000 in work around and inside the plants. Of these 10,000 labourers, 6000 were Chinese. "The process of canning (making cans, filling, cooking, soldering and boxing) is almost exclusively done by contract. The contracts are made with boss Chinamen who hire their own help in their own way" (135).

The advantages to the canners are: First, the contractor takes the responsibility of employing sufficient hands to do the work, thereby saving all the inconvenience and trouble which would otherwise fall upon the employer; second, the work is done by experts who have been trained to the business; third, the canner knows exactly what "the processin" will cost per case; fourth, any loss falls upon the contractor; fifth, he avoids the trouble of furnishing supplies, and the expenses of providing accommodation suitable for white men; sixth, the Chinese boss is able to get more work out of the men and to have it done more satisfactorily than when they work by the day for the cannery employer. (RC 1902, 135)

By the turn of the century, most canners admitted that if they were to hire white men to do this work, they would have to train them. Not only had the Chinese labour force become skilled, it was also scarce. The Automatic Can Company, located in New Westminster and employing white workers (many of them boys), supplied only about a tenth of the cans used on the Fraser River. William Campbell, company manager, explained:

The factory running to its full capacity will turn out about 200,000 cans a day. We employ from thirty to eighty, all whites, men, boys and girls. We make cans for some of the canneries ... I believe we can make them more cheaply than the canneries can make by hand. I believe that the labour is about one-half between the factory and hand labour. We have been in existence four years. The labour costs about $1 per thousand cans. The canners say they have to have Chinese help, and they employ them in making cans for some time before the fishing season so that they will have the men on hand when the fishing season comes on. (RC 1902, 154–5)

Because they were increasingly short of an abundant supply of cheap labour, canners also began to mechanize their lines. By 1900 machines had displaced an estimated 50 percent of the total labour force. However, the impact of mechanization was softened for a few years due to the phenomenal salmon runs and fierce competition to enter the industry by constructing new salmon canneries. This lasted until 1902 when B.C. Packers was formed, resulting in the closure of a large number of plants.

By using machines, canners could lower the contract price per case. But wages paid to individual Chinese labourers continued to rise. B.C. canners faced further competition for this supply of labour, as well as having their salmon runs intercepted, when salmon canning began on Puget Sound (testimony of Mar Chan, Chinese contractor, RC 1902, 141–2).

It was a Fraser River canner who, in 1891, began operating the first cannery on Puget Sound. A number of Fraser River canners followed, seeking to escape the ever more stringent Canadian fishery regulations. In the end, when U.S. canners took control, the process proved to be self-defeating. By 1896 there were fourteen canneries in the area. While it was illegal to use traps on the Fraser, there were no such restrictions on Puget Sound. This was a cost-saving measure since it meant that American canners did not have to hire fishers. Fish could also be kept alive in the traps for several days, thus allowing canners to spread processing over a longer period. In addition, operators on Puget Sound had access to the American market (there was demand in the south for cheaper grades of canned salmon), while B.C. canners depended heavily on overseas markets demanding quality fish and faced severe competition in supplying these markets. Since there was a demand in the United States for the cheaper fall species, a longer processing season resulted. This in turn lowered production costs, including machinery, and allowed Puget Sound canners to put up a cheaper pack than Fraser River plants. And a longer packing season provided a longer period of employment, making the southern plants more attractive to Chinese crews. U.S. canners also employed the contract system with overall labour costs 20 percent lower than on the Fraser River. The longer season allowed contractors to bid for lower prices per case packed. On Puget Sound, canneries generally operated from the first of May to the end of November, while on the Fraser the average season lasted only from four to eight weeks.

U.S. plants enjoyed yet another advantage in that two-thirds were located close to urban centres, allowing canners access to yet another supply of cheap labour power – white women and children. Since Fraser River canneries were located seven to twelve miles from cities, most of the labour force had to live on site. After hearing all the testimony, the commissioners concluded that while wages were higher than in the past, the proportion of the total wage payroll going to Chinese labourers was still very low. They also concluded that there were enough Chinese men available to work in the salmon canneries and thus further restrictions on immigration would not harm the industry (RC 1902, 165–7).

By the turn of the century, British Columbia was no longer a frontier society. Europeans were settling in ever larger numbers, and the urban population had begun to grow rapidly. Vancouver was incorporated as a city in 1886 with only 2000 inhabitants. In 1888 its population had grown to 8000 and escalated to 178,657 by 1901 (Adachi 1976, 38). A new urban labour force – white women and children – was increasingly available to do the work previously performed by aboriginal and Chinese labourers. But the racial categorization of jobs, which had allowed employers to hire cheap labour power, now acted as an impediment. "The occupations which usually afford work for boys, girls and women are all occupied to a great extent by Chinese and Japanese, with the result that steady employment is largely closed to the youth of the country and to women who have to seek employment of some kind to earn their living" (RC 1902, 211).

It is a bitter irony that soon after the United States forced China to agree to the free emigration of its population, U.S. authorities were attempting to stop it. In 1881 a treaty was ratified between the two countries allowing the U.S. to limit the number of labourers entering from China. In 1884 emigration of Chinese labourers to the U.S. was suspended for ten years, and in 1892 a further ten-year term was added. In 1892 resident Chinese were forced to register with American authorities, who could then monitor and control the movement of the resident Chinese population (RC 1902, 249–50).

In British Columbia, the construction of the Canadian Pacific Railroad required a large number of Chinese labourers. Once it was completed, however, a series of head taxes was imposed on Chinese immigrants: $50 per immigrant in 1885, $100 in 1900, and $500 in 1903. In 1875 Chinese residents were denied the provincial vote. Aboriginal peoples were also disenfranchised. The Japanese joined this group in 1895. It was only after the end of the Second World War that each group regained the right to vote. While all these measures were designed to curtail immigration, the Chinese Immigration Act of 1923 (also known as the Exclusion Act) stopped it altogether (Bolaria and Li 1985, 86–7, 106). The provincial government had been actively passing legislation to exclude the Chinese since the 1880s, but since China was an ally of the Western powers during this time, foreign diplomacy forced the federal government to disallow most of the provincial legislation.

TECHNOLOGY AND LABOUR (1905–12)

When labour was plentiful and cheap, machines were used to supplement the manual process. For example, gang knives were used to speed the cutting process, while retorts modernized the cooking process. "Prior to the introduction of butchering machines, the speed of the line depended on the capacity of the gangs to produce butchered fish, and there was little point in mechanizing other stages of the canning line, even where the technology was already developed" (Stacey 1982, 21). When labour became scarce and costly, efforts were made to mechanize the lines.

The iron butcher was introduced in the early 1900s and was sufficiently developed by 1907 to automatically clean the fish and supply two or three canning lines. This machine was the only one developed specifically for the salmon-canning industry. It became known as the "Iron Chink" because it displaced skilled Chinese butchers and there are iron butchers still in existence that bear the term on their patent plates. "It was no mere coincidence that the Iron Chink was installed in the Fraser River canneries immediately after the first labour shortage between 1902 and 1905" (Stacey 1982, 21). The butchers, whether human or mechanical, fed the canning lines. Mechanization increased the speed and regulated the quantity of fish being processed. As more machinery was introduced, more canning lines could run simultaneously, and overall output increased.

Although the iron butcher displaced skilled workers in one area, it and other machines obliged the canners to employ skilled workers for other purposes. Machines required maintenance, especially in the wet and dirty conditions of salmon canneries. In addition, since machines were operated for only a few months, they had to be overhauled and prepared for each canning season. Unlike other workers, these machinemen (most of them white men) have always been hired directly onto company payrolls, usually on a permanent basis.

By 1912 the sanitary can and double seamer were added to the line. With the Smith butchering machine (a further refinement of the iron butcher), two operators performed the work of fifty-one expert Chinese butchers (*SP*, no. 22, 1907, xli, lxii). Stacey (1982, 23) estimates that the adoption of the sanitary can and double seamer led to a 30 to 35 percent reduction in the labour force. American Can Company introduced them in British Columbia and was able to attain a virtual monopoly over this type of can-making machinery. It leased most of the machines on the lines to the canneries, and when many canneries ceased their local can-making operations, American Can became the chief supplier. Local operations continued in the north because it was costly to ship empty and bulky cans to remote canneries. Collapsed cans did not prove feasible, and preformed cans became standard in the province.

Thus a series of machines was introduced at various stages to overcome bottlenecks and labour shortages. These machines were usually fed by other machines interconnected by conveyor belts. Workers monitored the process and maintained, repaired, and adjusted the machinery. The machinemen, many of whom held engineering tickets, became ever more important. Yet until recently, canning operations have employed large numbers of manual workers to wash the fish being fed into iron butchers as well as to fill cans. Aboriginal women provided the original labour; later, women from other ethnic groups joined them. The vast majority of these women have been recent immigrants, their job opportunities limited by their inability to speak English. They took the place of Chinese men as a reservoir of cheap wage labour (Muszynski 1987, 62–4).

CONTRACTING THE LABOUR OF JAPANESE FISHERS

The white population of British Columbia at the turn of the century made few distinctions between Chinese and Japanese residents, lumping them together as "orientals" or "asiatics" – distinguished from "whites" (another racial category current in the popular imagination and in the local press) because they accepted wages upon which the latter could not survive. Nevertheless, when the Japanese began to immigrate in large numbers in the late nineteenth century, Chinese labourers perceived them as a threat because Japanese contractors undercut the wages of the Chinese, not only those of white male workers. They did not compete with the Chinese in the fishing industry, however, since the original Japanese immigrants were fishers and continued in this line of work. It was not long before they began to send to Japan for brides, and Japanese women were working inside the fish plants. But they took jobs already designated as "women's work," jobs like filleting (fresh fish began to be processed within salmon canneries with the adoption of refrigeration) and manually filling cans, both paid on a piece rate basis. Aboriginal women working on the Fraser River appear to have avoided these types of jobs, preferring to be paid hourly wages for jobs such as washing fish (Sparrow 1976, 132).

While the Chinese were described as a "feminine" race and judged to be taking jobs away from white women and children, thus hindering the establishment of working-class families in British Columbia, the Japanese were referred to as a "manly" race. Joseph D. Graham, government agent based in Atlin, testified: "The Japanese are a little cheaper than the Chinese. I would rather deal with them. They are a more manly class of people. They purchase goods from our ordinary tradesmen. They have more of the western method about them. Everybody has his own idea. I draw my own conclusions from what I have seen of the Japanese: that they are a more manly race of people, and I have always drawn that conclusion. I have only met a few of the Japanese. Those I have met have been more manly than the Chinese. I cannot speak of them as a race." Clive P. Wolley, former executive officer of the B.C. sanitary commission, described the Japanese as willing to "live as a white man does," as more likely to assimilate, and as more "manly and gentlemanly." But he is therefore "a more dangerous competitor with the white man": "He adapts himself more easily to our civilization than the Chinese. The Chinese will do the lowest kind of labour and stick to it. The Japanese will get higher if he can, and he has brains enough to rise into any of the mechanical pursuits" (RC 1902, 375–6).

Not only did the Japanese enter the fishing industry as fishers, they took up the boat building trade and also competed with white male labour in jobs that supplemented fishing during the winter months. Charles Kilby, a fisher living in Nanaimo who fished the Fraser for many years, testified: "They are also doing the work of the white fishermen which we used to do during the winter months, when the salmon fishing was over. The white fishermen used to supply cordwood to the canneries, and clear the land and cut shingle bolts. The Japanese do that now, and in fact they have almost monopolized the unskilled labour that the white fisherman used to work at during the winter" (RC 1902, 343). Especially interesting in the white fishers' testimony to the Royal Commission was the frequent mention of their families. For example, John Kendall, of New Westminster, told the commissioners: "I am a Newfoundlander by birth; am married and have five children, four boys and one girl, oldest 12. I am over $200 behind what I was last year this time. If the Japanese and Chinese still continue to come I have got to leave or starve. I am British to the backbone. I wish to stay under the British flag" (344). Kendall claimed he was being undercut by the Chinese and Japanese not only in fishing but also in the sawmills and factories.

Nor was it only the "white" fishers who were protesting. A section of this part of the commission's report is titled "Protests of Indian Chiefs." The testimony of Chief James Harry, of the Squamish Indians, representing seven Indian chiefs, was typical if more detailed than the rest:

The Japanese are cutting all the wood we have here in British Columbia and bolts on the north arm, Howe Sound, and here in Vancouver on the west side, and our people have no chance to go to work and cut the cedar. They used to cut the cedar and bolts and wood. The Japanese cut wood for too little – just like for nothing. My people worked in the mills; now have no chance. The Japanese work for about $15 a month – not enough to buy clothes and keep wife and family. I have a wife and three children. Thirteen and fourteen years ago the Indians got $1.50 and $1.25 working in the mills; now they get no chance to go to work. The Japanese can live on a tablespoonful of rice and a little perch. We are not the same. I think the Indians and whites as good as the Japanese. The Japanese build boats cheap and make oars. We make sails, boats and oars and everything. Our women get work in the canneries; they get $1, $1.25, $1.15, $1.10, depends on what they do. Boys and girls get work when plenty of fish. We do hand-logging in winter. We do stevedoring, make good wages but not steady. I think you could get enough whites and Indians to do the work. There are a little more than 10,000 Indians, men, women and children, engaged in fishing. (RC 1902, 346)

Chief Harry's testimony is worth quoting at length because it not only demonstrates the extent to which the aboriginal economies had become intertwined and dependent upon capitalist wage relations but it also demonstrates the complexity of the intertwining of race and labour. The competition for jobs was phrased in racial terms but it was also clear that the ability of any one group, or "race," to undermine the wages of another depended on the ability of that group to survive at the lowest possible wages, and wages were determined, in the last instance, by capitalist employers. However, employers were also constrained and had to pay wages that allowed the various groups to subsist, however meager the level of subsistence might be, and employers were also faced with organized groups who pressed for collective bargaining over fish prices, wages, and working conditions. In the 1890s and the early years of the twentieth century, Japanese contractors or bosses approached canners and other employers, offering to guarantee the supply of a sufficient labour force. In the overcrowded conditions of the fishing industry during these years (overcrowded in the number of canneries and number of boats fishing), employers stood to gain by agreeing to these contracts. Alexander Ewen, pioneer canner engaged in the cannery business since 1870, noted: "I don't think the number of canneries would be here for the cheap labour. The canners think there are too many canneries and the fishermen think there are too many fishermen. The white fishermen have dropped away. The Japanese are taking their place" (RC 1902,

341). Fishers had begun to organize collectively and the years 1900 and 1901 were marked by labour unrest on the fishing grounds (Ralston 1965). Japanese fishers were employed not only to hold down the price of fish but also as strikebreakers, which created a protracted and intense level of racial tension among Japanese, aboriginal, and white fishers which culminated in the evacuation of the Japanese from coastal British Columbia and the confiscation of their fishing vessels. Prior to the Second World War, as a result of yet another fisheries commission, the federal fisheries department implemented a policy of eliminating Japanese fishers. Between 1922 and 1925 over one thousand men were driven from the industry (Meggs 1991, 127).

> In 1921, Motherwell's officials had licensed 6,229 gillnet and troll fishermen coastwide. Of that number, about 1,200 were native people, of whom 969 gillnetted for northern canneries. Another 2,200 were white, including 1,600 gillnetters: again, the majority – some 868 – gillnetted north of the Fraser. The Japanese Canadians made up more than one-third of the fleet, operating 2,600 boats; of these, 1,100 were working in northern waters, another 873 on the Fraser and almost all the rest were Vancouver Island west coast trollers. According to Found's information, the total Japanese-Canadian population of the province was only 15,000 men, women and children, of whom 9,800 were male ...
>
> Driven by contractors and economic necessity, the Japanese-Canadian gillnetters consistently outproduced their white and native counterparts and were cheaper to employ. Three northern canneries reported that in 1922 the average Japanese-Canadian fisherman landed 5,451 fish on the Skeena. Native fishermen landed 3,220 and whites 3,192. The reasons lay not in any genetic superiority but in the fact that Japanese fishermen started the season earlier, often were given better gear and worked together to mend nets and equipment. (Meggs 1991, 126–7)

To understand the dynamics of the transformation of racial tension into racism, the forced removal of an entire "race" of people from the fishing industry, we must first understand how it was that the Japanese could undercut the earnings of white, aboriginal, and even Chinese labourers. Adachi's account in *The Enemy That Never Was* of the events in Japan that led to emigration bear many similarities with what had happened in China (1976, 1–35). The Portuguese also visited Japan in the sixteenth century, another country that had closed its borders to foreign contact for several centuries. The Togugawa Shogunate came to power in 1603, and the ruling class of the times also held merchant activity in low esteem. But it was the religious rather than the mercantile influence of the foreigners that the rulers feared.

By 1600, out of an estimated population of between fifteen and twenty million, some three hundred thousand had become Catholic thanks to the zeal of Portuguese and Spanish missionaries. A 1614 edict banning Christianity and expelling the priests was followed, in 1624, by a ban on Spanish ships and traders. The law was further tightened in 1638, when only Dutch traders were allowed entry and confined to a small island. In 1637 any Japanese caught trying to leave or to re-enter the country faced the death penalty. Some of the first Japanese to land on the B.C. coast did so inadvertently, their small boats having been shipwrecked in the vicinity (an upper limit on the tonnage allowed in boat construction, another measure to limit the distances that could be travelled, made for large numbers of losses at sea).

A period of peace and stability, witnessed by a growth in the population, followed this imposed isolation. But this period ended abruptly in 1853, the year Commodore Perry anchored a squadron at the mouth of Yedo Bay, forcing Japan into the Treaty of Kanagawa. U.S. ships were henceforth to be allowed to trade at two ports, and a consul was to be appointed. Similar agreements were made with the British, Russians, and Dutch. In 1867 the Togugawa Shogunate fell and the Meiji emperor regained control of the state, embarking on a period of rapid industrialization and westernization. Emigration, however, was still outlawed. In 1868, 153 Japanese were pirated to the Hawaiian Islands, an activity that resembled the Chinese coolie trade.

The Japanese state played a more direct role in emigration than did the Chinese. In 1884 Japan signed a convention with Hawaiian sugar plantation owners to import Japanese labourers under contract. Emigration was legalized soon thereafter, and in 1885, 943 Japanese arrived in Honolulu. Over the next nine years, 28,000 labourers on three-year employment contracts reached Hawaii. By 1908, 178,927 of them had entered the Hawaiian Islands:

> In 1884 the Japanese government signed a convention with the Hawaiian government whereby the sugar plantation owners in Hawaii were permitted to import Japanese labourers under contract between 1885 and 1894, and soon after that a law to permit general emigration of labourers was enacted. Thus overseas migration began with the 943 farmers as government-sponsored emigrants to Hawaii. The fact that they were chosen among 28,000 applicants shows that the impoverished conditions of rural communities pushed the farmers to hope for a better life abroad. They were the first of over 29,000 to enter Hawaii on three-year contracts between 1885 and 1894, and the first of a total of 178,927 who entered Hawaii before 1908. They were also the vanguard of emigration to the United States, Canada and South America, where 91,740 Japanese went before 1908. Also some of those who went to

Hawaii eventually came to Canada, and the fact that their number increased in 1906 and 1907 was to become one of the causes of the Vancouver Riot. (Iino, 1984, 30–1)

In 1877, Manzo Nagano, a sailor, arrived in British Columbia. Together with an Italian partner he took up fishing on the Fraser River. Word went out and in 1884, thirteen more men came and settled in New Westminster and Steveston. Immigration then began in earnest.

As in China, most immigrants came from the peasant class and from the same area in Japan. The large community of Japanese fishers that was soon established in Steveston came from the small village of Mio in the prefecture of Wakayama, where peasants and fishers were having a very hard time subsisting with labour productivity approximately half the national average. A rocky coastline and stormy weather made fishing difficult and hazardous, and the granting of offshore fishing rights to a neighbouring area rendered seventy fishing boats in Mio inactive. In 1887 a carpenter from Mio visited Steveston and was soon followed by most of the young men from his village. Mio became known as "America-mura" and prospered thanks to the remittances received from Canada. If they had the means, most young fishers returned to Mio during the winter season (Adachi 1976, 18–19).

A similar series of events in both China and Japan created a large surplus population with little means of subsistence to keep from starving. As Japan was drawn into the international economy, it embarked on a program of rapid industrialization – financed on the backs of the peasantry. Rice remained the standard of value, and landlords drained the peasantry of its produce. Taxation levels were astronomical, and infanticide and abortion rates rose in these years, testifying to the inability of many families to feed all their members. All this ended in a series of agrarian revolts. Adachi counts at least 190 in the decade following 1867. Industrialization was imposed on a basically feudal economy which could not sustain the pace. As late as 1885, 70 percent of the population subsisted on agriculture. Unable to pay taxes, many peasants were forced to sell their land. They became tenant farmers or left the land to search for employment in the urban areas. But the population had escalated from thirty-five million in 1872 to forty-four million in 1900, and the towns could not possibly absorb this surplus. Many chose emigration as a solution.

As in China, the merchant class, despised in feudal Japan, acquired power when the state began to promote industrial development. Merchant companies sprang up to handle contracts to supply overseas demand for labour. Although the government tried to maintain control by limiting emigration, these firms acquired effective control. An emigration company had to apply to the Japanese foreign office for permission to send out a specified number of labourers. Each labourer had to supply a passport and a certified copy of his family register and to undergo scrutiny by the local police. Labourers were expected to return after a specified period of time. Companies that came to specialize in labour contracting were originally formed to meet the demand for cheap labour on the Hawaiian sugar plantations. In 1884 the Japanese government, in an attempt to prevent abuses, stipulated that these "agents" had to be incorporated under the authority of the minister of home affairs. The companies continued to grow in number and size, and so apparently did abuses of the system. In 1900 there were twelve companies which had grown to seventeen by 1907. The following year, in 1908, all but three of these were suppressed (Adachi 1976, 25).

Like their Chinese counterparts, Japanese firms were in the business to make a profit. They frequently induced peasants and labourers to sell their land and belongings. If the emigrant lacked sufficient money for the fare, the firm advanced him a loan. A commission of up to 25 yen per head was collected for services such as assistance abroad and a guaranteed return trip home if the labourer became sick or impoverished. "It was claimed during the Royal Commission proceedings of 1900 that emigration companies, agents of steamship companies, brokers and boarding house keepers were all intimately connected in an unbroken chain to exploit the situation. If to the list were added the contractors and the gang 'bosses,' operating in Canada to lead the men to work, then the 'organized scheme' was further extended" (Adachi 1976, 25–6; RC 1902, 330–6).

Adachi notes that by 1900 Japanese labourers were displacing Chinese labourers in such occupations as railroad construction. Roy (1989, 205–6) notes, for example, that Saori Gotoh and Frederick Yoshy of the Canadian Nippon Supply Company held a near monopoly in supplying Japanese labour in British Columbia; and, in October 1907, the company had 1468 labourers under contract, most of them employed by the Canadian Pacific Railway Company, which also carried many "Asian immigrants" on its trans-Pacific shipping line. According to Adachi (1976, 32), the men were young, single and knew little English (although their level of education was generally higher than that of their Chinese counterparts). The Japanese contractor in Canada acted as an intermediary between the employee and the employer. Although the immigrants originally worked individually, when their numbers began to rise, Japanese contractors would organize groups of men and negotiate contracts on their behalf, "assuming upon himself the expense of maintaining them, all the while retaining for himself a profit on the transaction" (31). A commission of 10 cents a day was charged for each day worked. The cheap boarding and lodging houses found in Vancouver's Japan town was a major recruiting ground for contractors. Adachi (32) notes: "Indeed in the early years, as the immigrants stepped from boat to lodging house to gang, there did not seem to have been a real break from their homeland."

If we return to the concept of use value developed in the earlier chapters, we could make a comparison between the contractor physically maintaining his Chinese and Japanese labourers and European men maintaining their wives and children in the nuclear household. The labour of the Chinese, the Japanese, and of women and children then becomes available as cheap labour, mediated through the contractor on the one hand and through the husband/father on the other. It is obviously much cheaper to maintain single Chinese and Japanese men than it is to reproduce the labour not only of the current generation of workers but also that of the next generation. In the case of contract labour, the next generation was being reproduced in China and Japan, societies in which women and children were also subordinated under the respective patriarchal relations of oppression.

Another difference is that white women were providing the work necessary to maintain the "free" male labourer, while Japanese and Chinese labourers did not maintain their contractors with direct work (use value) but indirectly through the profits contractors were able to make (often in the form of absolute surplus value, since contractors could frequently only make a profit by cutting back on food and working their labour force all that much harder). The two sets of situations contain many obvious differences, but when one considers them in the light of the capitalist's need for cheap labour, striking similarities begin to emerge. And it is important to keep in mind that the labour contract lasted only as long as the men had no families with them in Canada. When Chinese and Japanese men began to bring wives to Canada, a different wage relationship with employers began to emerge. Their wives now supported them, not the contractor. This allowed the men to exercise direct control over the labour of others. And so their own wives and children became available as a cheap labour force, demonstrated in the salmon canneries with the entry of Chinese and Japanese women early in the twentieth century.

The price the labour of these women commanded undercut that of white women because their households, like those of aboriginal women, preserved some of the pre-capitalist relations of production found in their countries of origin. For example, the Chinese and Japanese established separate communities (often they had no choice given racist exclusion) with extended family households. Older people were available to look after the children, and a number of people earning low wages could collectively contribute a decent income to the household.

Most Chinese women working in the fishing industry in the 1980s were to be found at Canfisco's Home Plant, located just on the boundaries of Vancouver's Chinese and Japanese communities, thus providing a ready and abundant supply of cheap wage labour. This recruitment pattern has been extended to other newly immigrating groups. For example, in the mid 1980s the Prince Rupert plants hired large numbers of Portuguese and East Indian women. They too lived in extended households, spoke little or no English and were willing to take any kind of work. Their husbands often worked in the sawmills and other resource industries and were subject to seasonal layoff, making it imperative that their wives and other family members find jobs, however menial. The seasonal nature of women's employment in the fishing industry is indicated by the fact that while women comprised over 50 percent of the seasonal plant labour force, during the off-season their percentage fell to as low as 5.3 (UFAWU 1984, table A, 28a). The permanent labour force was overwhelmingly male (and white).

Picking up the history of the contract system for Japanese labour, by the turn of the century, the Japanese contractor often had a wife to help him. She usually did the cooking and the laundry for the labour gangs (Knight and Koizumi 1976, 30). One man recalls that, when he married a picture bride in 1906, the canner promoted him to become the boss of a boarding house (probably a bunkhouse), placing him in charge of forty to fifty men: "The 'boss' earned a commission according to the catch of his 'boys' and negotiated with the company on their behalf, while his wife cooked and served as a mother-substitute for the 'boys'" (Marlatt and Koizumi 1975, 11). The boss also controlled between twenty and thirty boats and received the advances made by the canners for the labour of the fishers.

Japanese women began to immigrate in larger numbers toward the end of the first decade of the twentieth century. Many were "picture brides" who had never seen the husbands they came to join. A man would write to his family and request that a marriage be arranged. The family would search for a suitable mate and send her picture to the prospective groom. Upon his approval, the marriage was registered in Japan and the marriage ceremony was repeated in Canada. Since these were years of tightened immigration control in Canada, this was often the only way Japanese women could enter the country if they had no relatives residing here. Many of these women initially helped their husbands on fish boats, as had aboriginal women (most boats were gillnetters requiring two fishers to handle them). But cannery bunkhouses were reserved only for the men (Skogan 1983, 83), so married men had to search for other accommodation. In the south this was relatively easy, since Steveston was developing as a fishing village with a large Japanese population. And the introduction of gasoline engines during this period meant that longer distances could be travelled in a shorter period of time, allowing fishers to commute from their homes in urban centres.

In May 1901 the commissioners investigating Chinese and Japanese immigration visited Steveston. They found a "busy hive of men" almost all of them (except supervisors) Chinese or Japanese. The Chinese were busy making cans while the Japanese were building and repairing boats for the coming season (RC 1902, 357). The number of Japanese immigrants had risen from 691 in the year from 1 July 1896 to 30 June 1897 and to 9033 in 1899–1900. (The latter figure was unusual. Most emigrants had been bound for Honolulu but the port there was closed due to an outbreak of bubonic plague. The United States was not accepting Japanese immigrants, and rather than return home, the vessels continued on to British Columbia.) The total number of immigrants from 1896 to 1901 was 13,913 (327).

Many of these immigrants entered the fishing industry. They began to acquire a large proportion of the fishing licences. While in 1896 they held only 452 of 3533, by 1901 they held 1958 of a total of 4722 licences. Each licenced fisher usually employed an unlicenced boat puller. Unlike white and aboriginal fishers, the Japanese fished under contract to the canners, and depended on the canners and contractors for boats, advances, housing, and provisions. They were a subordinated labour force, unlike independent white and aboriginal fishers.

Over the years, increasing numbers of white fishers entered the industry and became a militant labour force, pressing canners for higher fish prices (at this time they were paid by the fish). In 1900 and 1901, when two major strikes occurred, the canners predictably used Japanese fishers as strike breakers (Ralston 1965). In fact, canners probably employed Japanese fishers as a means of controlling fishing on the Fraser after limits on licences were lifted, allowing a flood of independent fishers into the industry. The racial hatred that has marred the history of the provincial fisheries dates from this period, and culminated in the forceful evacuation of Japanese residents from the coast after the bombing of Pearl Harbour. By that time the Japanese had become known for the quality of their boats, for possession of which fishers disgracefully battled one another.

In their testimony to the Royal Commission (1902, 340–2), canners argued that Japanese fishers were indispensable for the same reason that Chinese cannery workers were essential: both groups provided cheap labour power. It could be argued that the Japanese were hired on the fishing side to mirror the earlier role of Chinese cannery crews. White fishers on the Fraser were becoming increasingly militant and organized and often able to form alliances with aboriginal fishers against the canning companies. At this point Japanese fishers as a subordinated labour force had little choice but to be used in breaking strikes. Later, when they too became independents, militancy among them rose. White, aboriginal, and Japanese fishers can all point to instances when one group was striking while another group broke their strike. There are also instances when all three groups acted cooperatively. Before the Second World War, however, racial categories marked the groups off from one another. Certainly each group was situated differently in its conditions of survival and in the fishing industry. But there can be no doubt that racism flourished during this period of time, and provided justifications for the actions of one group based on colour of skin. Clearly, whites dominated in all respects.

When the Japanese were allowed to return in 1949, racist hostility in the industry had become muted. But the Japanese who returned to fishing no longer labourered for contractors. In fact, the entire pattern of emigration from China and Japan had changed. Iino (1984, 29) divides the history of Japanese immigration to Canada into four periods: 1877 to the Lemieux Agreement of 1908; between 1908 and the outbreak of the Second World War; evacuation during the war years up to 1949; and from 1950 to the present when "Japanese-Canadians gained full rights as citizens of Canada." During the fourth period, professionals with university degrees and valued skills were entering Canada and changing the Chinese and Japanese communities already established in the major urban centres. They no longer settled primarily in British Columbia, and very few were interested in such menial jobs as those in the fishing industry. It was left to new waves of immigrants from other countries to take their places. But the industry itself had also changed dramatically. Federal labour legislation and the establishment of an industry-wide trade union in 1945 were to have profound effects on shoreworkers, and these form the topic of the following chapter.

5 Organized Resistance: The United Fishermen and Allied Workers Union

Discussion of cheap wage labour has thus far centred on capitalist exploitation of the fisheries and the salmon canners' efforts to hire labour as cheaply as possible, efforts which used pre-capitalist relations of production within British Columbia and elsewhere to devalue labour power through categorizing people by gender, race, and ethnicity. Although those whose labour power was devalued resisted, resistance tended to be based in relations outside the workplace. There are instances of struggle over wages and working conditions within the canneries, but until the period following the end of the Second World War, these tended to be isolated and localized. Fishers provided the impetus to unionize, and they needed shoreworkers to create an industry-wide trade union. In the early period shoreworkers tended to accept negotiations dominated by the demands of fishers, demands which centred on negotiations over the price of fish. But shoreworkers gradually began to assert their own needs and interests. Part of the struggle involved trying to abolish distinctions based on gender and race to keep wages depressed for a majority of the plant labour forces, distinctions that were maintained in early union agreements, as is evident in the supplements found in the appendix. This chapter examines events that mainly transpired after the Second World War.

Early in the fishing industry, fishers had a reputation for being far more militant than shoreworkers. Only when the United Fishermen and Allied Workers Union had organized to include *all* shoreworkers in its contracts did plant workers begin to press for concerns specific to their own working conditions. However, it is misleading to assume that those who laboured for wages were totally passive. For example, Chinese labourers organized the Chinese Cannery Employees' Union in 1904 to deal with contractors who left for China after being paid by cannery operators without first paying their own crews (Gladstone 1959, 296–7). And Meggs (1991, 69) records several instances of plant disputes in 1901, the year of the big fishers' strike and the year of the big sockeye run:

> With the run surging past, the Chinese workers could assert their authority. At the Gulf of Georgia cannery, a foreman was surrounded by his crew and agreed at knifepoint to rehire two contract workers he had fired for sloppy work. A few weeks later another crew in New Westminster threw down their knives in protest against the brutal pace set by an iron butcher and automated canner. In a third strike, the *Province* reported on 4 September that Chinese crews halted for a time the installation of a fish-packing machine which "takes the place easily of about 30 Chinese." The machine "worked so fast that it took the Chinese all their time to keep up with it and they had to hustle. Finally they threw down their knives and stated that either the machine must go or they would walk out of the factory. The machine was shut down." On 4 October the *World* reported that a foreman had to draw his gun to subdue a crew that threatened him with knives. He "warned the foreman of the Celestial gang that he would shoot the first man who disputed his authority."

The extensive use of the Chinese contract system made it effectively impossible for those shoreworkers hired by contractors to bargain directly with processors. This was one of the chief advantages to employers when they used middlemen contractors to hire, feed, and pay the "China gangs," and when they adapted the contract system to hire seasonal labour in general. The proportion of inside plant workers hired on a permanent basis and paid directly by the canners was tiny. These were predominantly white men, generally hired as managers and to maintain the plant equipment. It is not surprising that the UFAWU initially recruited these men when it began to organize shoreworkers, and was reluctant at first to include all seasonal labour in negotiating contracts.

Fishers had a history of militant organization that went back to the nineteenth century (Gladstone 1959; Ralston 1965). Unfortunately, racial segregation divided fishers and prevented them from uniting collectively in their negotiations with canners. While the racist climate in British Columbia was a source of conflict, there were also profound divisions among groups of fishers that related to their different economic, linguistic, cultural, and political backgrounds. Thus European fishers formed union organizations that replicated a tradition of organizing on the basis of craft skills, and created a number of unions and associations based on the type of gear used (gillnet, seine, hook and line for trolling; later expanded to include combinations of gears and of fisheries). Barman (1986, 41) notes: "Fellow immigrants from Britain were appraised according to British notions of class, with traditional perceptions being reinforced by the leading role of British working-class immigrants in trade unionism in British Columbia."

Ethnicity and racial and gender categorizations formed part of the "ethos" of working-class trade unionists as well as that of the ruling classes. Class interests also proved divisive among fishers, some of whom were able to earn substantial incomes and began operating as small-scale capitalists, employing boat crews as wage labourers or on a share basis. Gear type was again a factor, as was the particular fishery that was exploited.

Aboriginal fishers sought to preserve their traditional right to use sites and capture fish for their own consumption, and the food fisheries were included in their land claims. The associations they formed, like the Native Brotherhood, reflected these wider interests, in which commercial fishing figured as only one component (Newell 1993).

It must also be kept in mind that working-class organizations like industrial trade unions arrived with capitalism in the Pacific Northwest. As Meggs documents in *Salmon: The Decline of the British Columbia Fishery* (1991), organized fishers have often seen their own claims to the resource threatened by aboriginal land claims that include the fisheries. There is significant competition between non-aboriginal fishers and aboriginal fishers over the same resources. For example, on 16 September 1994 *Globe and Mail* reporter Miro Cernetig wrote that 1.3 million salmon were "missing in the Fraser River ... Two years ago, when about half a million fish disappeared, many accused natives of poaching. But Mr Chamut [assistant deputy fisheries minister] said that neither poaching nor inaccurate counting by biologists of the fish arriving at spawning grounds could account for this year's shortfall." Having exonerated aboriginal fishers one week, the following week another article appeared laying the blame at their door. Cernetig (1994b, A1) reviewed a seven-page memorandum dated 17 August 1994 and concluded that "the ability to properly manage the Fraser River Aboriginal Fishery has been seriously compromised." The document "makes no estimate of the number of fish that may have been lost to illegal fishing activities. Most experts believe it is unlikely that illegal fishing could account for all the more than 1.3 million missing fish" (Cernetig 1994b, A1). Yet the article links aboriginal fishing with "illegal fishing activities" and cites the memorandum's conclusion that the Fraser River Aboriginal Fishery is improperly managed. By whom? The department of fisheries and oceans. Almost a century and a half after its creation there is still little evidence that the federal government has listened to the First Nations and their insistence on recognizing their aboriginal rights to the fisheries as other than something that has to be "managed." This is still a central area of dispute between non-aboriginal and aboriginal fishers since the two groups have directly competing claims to the fisheries. Non-aboriginal fishers are directly tied to successful capitalist prosecution of the fisheries as an industry, and unionization reflects their struggles on this terrain. Aboriginal fishers thus have a contradictory place in union history since they occupy more than one terrain. And aboriginal shoreworkers cut into and across all these terrains in yet other ways, as proletarianized workers. The history of their place in the UFAWU is instructive in revealing some of these conflicts and contradictions.

As mentioned in chapter 4, Japanese fishers originally worked on contract and were used by canners as an alternative source of supplying the raw resource to undermine the bargaining strength of European and aboriginal fishers. In the north, canners also tried to employ aboriginal fishers as wage labourers, using licence limitation to control aboriginal fishers who refused to take out licences under their own names, resisting the implied acknowledgment of capitalist control and state regulation of fisheries they considered to be their inheritance (BCFC 1892, 15). In summary, until the end of the Second World War, the industry was marked by conflicts among fishers reflecting not only the prevailing racist ideology but also structural barriers that situated each group differently in the capture of the resource, in its relationship with processing companies, and toward one another.

Despite all these internal conflicts, fishers were generally far more militant than shoreworkers. They enjoyed a measure of independence in negotiating directly each season with canners over the price of fish, and most of the struggles shoreworkers initially entered were to support fishers in their demands. This was especially noticeable when fishers and cannery crews were directly related; for example, when the two groups came from the same villages. As mentioned in the previous chapter, in the 1930s fishers became more militant, and there was increased militancy among shoreworkers. But they continued to support the demands of fishers. If the fishers succeeded, shoreworkers occasionally received concessions for themselves as well.

The depression years witnessed an escalation in labour militancy throughout Canada. In the B.C. fisheries, union organizers affiliated with the Communist Party undertook with fishers to establish a province-wide industrial union. To achieve union status required a membership of wage labourers, and thus shoreworkers were crucial to their success. In addition, by unionizing all shore plants, fishers were assured that canners could not process catches from non-union boats. Indeed, in the early period of the UFAWU, the interests of fishers predominated. Shoreworkers had to learn how to put forward issues of concern to plant workers within the structure of the UFAWU, and this required that they develop a class consciousness of themselves as wage labourers with issues pertinent to their own working conditions.

FISHERS ORGANIZE SHOREWORKERS: 1931–45

As noted, shoreworkers were a relatively unorganized labour force, compared to fishers who had established a number of strong craft unions. The UFAWU was the product of a number of earlier fishers' unions. Following a strike in 1931 in Barkley Sound by unorganized salmon seiners and gillnetters, the Fishermen's and Cannery Workers' Industrial Union was formed. In 1935 the union signed fish price and cannery agreements with the Deep Bay Fishing and Packing Company for the Deep Bay cannery located on Vancouver Island, but the agreements terminated the following year with the demise of the union. It was reorganized in 1936, and two separate unions emerged, the Salmon Purse Seiners Union of the Pacific (SPSU) and the Pacific Coast Fishermen's Union (PCFU) (North 1974, 9–22). In 1937, the SPSU and the PCFU jointly founded the union paper *The Fisherman*, and it became the organ for the UFAWU upon its formation. Much of the information in this chapter is the product of a content analysis of the issues covering 1937 to 1980.

In 1940 the SPSU joined the United Fishermen's Federal Union (UFFU) (*The Fisherman* [hereafter *TF*] 26 March 1940, 1). The UFFU was itself the product of a number of unions that underwent several name changes in the 1930s. In 1941 the PCFU merged with the UFFU (1941, 25 March, 1), and the UFFU helped to found the Fish Cannery, Reduction Plant and Allied Workers' Union, or Local 89 as it came to be called. The UFFU shared office space with Local 89 and provided financial assistance. The two were unable to merge because the Trades and Labour Congress of Canada insisted on a separate charter before it would grant jurisdiction for the organization of shoreworkers (1941, 15 July, 1). Both unions were affiliated with the Vancouver Trades and Labour Council (VTLC). The Trades and Labour Congress finally agreed to grant a provincial charter in 1945, when the two unions merged to become the United Fishermen and Allied Workers' Union (UFAWU) (VTLC 1945, 20 March, 312).

The impetus for organizing shoreworkers came from fishers. In the late 1930s they were interested in aiding shore plants to organize because it would give them added leverage in negotiating fish prices. Fishers needed shoreworkers because the key to shutting down the industry lay in closing down the plants. Otherwise, plants could operate by processing catches bought from non-unionized Canadian fishers and from American boats. However, a handful of key employees could close the plants: reduction plant workers and cannery machinemen. These two groups consisted primarily of European men whose interests were similar to those of unionized fishers (a number of fishers alternated employment between shore plants and fishing).

Bill Gateman, the shoreworker organizer for Local 89, recruited union members by job group, and thus union agreements came to reflect the racial and gender divisions prevalent in the industry. This recruitment pattern appears to have been common in industries marked by a division of labour using race and gender criteria and typified Gateman's experience. Previous to his appointment to Local 89, he had been the business agent for Local 28 of the Hotel, Restaurant and Culinary Employees and Bartenders Union (HRCEBU). There too the major portion of organized labour consisted of a small core of white males, waiters and bus boys (bartenders also appear to have been mostly European men). Beginning with this core group, organizers then attempted to recruit waitresses and kitchen workers (many of whom were Chinese), apparently with limited success. For example, Gateman noted in his report of the 17 July 1938 meeting: "Restaurant Owners oppose the inclusion of cooks in the agreement ... A motion made to take a secret vote on whether the members are in favor of accepting the agreement without the Cooks. Result of ballots showed 29 Yes, 18 No." A motion was unanimously carried to sign the agreement, with a recommendation that every effort be made "by all members to organising the Chinese" (HRCEBU, 139).

Earlier that year a delegation of Local 31 (Japanese Camp and Mill Workers), a union composed entirely of Japanese labourers, had approached Local 28. Both locals were affiliated with the VTLC, and Local 31 requested Local 28 to include Japanese employees in its new agreements. "After discussing the whole matter in a frank way, the Executive decided that our policy is that we cannot guarantee employment for Japanese employees in the new [Vancouver] Hotel" (HRCEBU 1938, 23 March, 120). Distinctions based on race were a common way of recruiting labourers to specific jobs in many industries that needed cheap wage labour, and they presented union organizers with particularly thorny and delicate problems in recruiting members in these industries. Take the following example: "It was pointed out by the Secretary ... that the [food] Canners in B.C. are the same companies which are operating in Quebec and Ontario. In the East they claim they are unable to pay higher wages on account of the cheap Oriental Labor in the West. In the West they claim they are unable to pay better wages because of the cheap labor in Quebec. This situation is to be watched by the Executive when being dealt with by the Provincial Minimum Wage Board" (VTLC 1938, 15 March, 242–3).

This is an example of the overlapping of categorization by ethnicity as well as race and again reflects the differing material circumstances of the two groups (Quebec francophones and "Oriental Labor") that force them to accept wages below those of the organized British male proletariat. This incident occurred in the 1930s. In the summer of 1984 I was driving from Vancouver to take a new position at the University of Regina and stopped for a meal at a small cafe in the southern Okanagan at the height of the fruit-picking season. A couple of young French Quebecers entered to use the washroom facilities. Their appearance provoked a vicious denunciation from the owner and the other men sitting in the cafe who were obviously regular patrons. The migrant fruit pickers were living in tents with minimal conveniences like decent bathroom and shower facilities (the reason the young people were using the cafe's bathrooms). But the men were castigating the workers themselves as somehow being filthy by nature and thus not caring for what to them would be "civilized" comforts requiring a higher standard of living. This was used to explain why they took lower wages. It shocked me that the same type of arguments used a century ago were once again being used against an ethnic group because that group was willing to labour for wages those who held some degree of power were unwilling to accept.

Early in 1938 there was also much discussion about the boycott of Japanese and "fascist" goods. For example, in the minutes of the VTLC meeting held 18 January 1938: "The Boycott Committee, delegate Shearer reported on the meeting held and pointed out that while conducting a campaign to buy no Japanese goods we do not desire in any way to injure the members of the Japanese Union affiliated with this Council and extended to them sympathies and condolence in these trying times" (233). At the meeting held 15 March 1938 the executive recommended making a statement "regarding the policy of the Council on the Japanese question. It should be distinctly understood that the Boycott of Japanese and Fascist goods is not intended in any way to be construed into a boycott against Japanese residents here. The matter of driving the Japanese Fishermen off the Fraser River and refusing to patronize Japanese stores is not the policy of this Council. The Boycott is against Imperialism in Japan." And under New Business a motion was carried "that this Council reiterate its previous stand as being opposed to Anti Oriental Ratial [*sic*] Prejudice campaign as being carried on by some in this City" (15 March 1938, 242). The Daily Press was mentioned and the motion passed 66 votes to 43, indicating considerable tension around this issue. In April 1940 there was a purge of communist members when the chair unseated three members (16 April 1940, 388).

In the pages of *The Fisherman* anti-Japanese hostility was also noted. In the issue of 2 September 1941 ("Cannery Organization Forges Ahead"): "For instance, the Japanese employees were told that only white people would be allowed to join the union; their objective being the elimination of Japanese working in the canneries. Again in other plants, workers were told that if the Japanese were taken into the union they would sabotage any united front in dealing with the employers" (2). And in the 7 October issue that same year the editorial was entitled "The Indian Problem": "Indians generally are low paid and as they are the bulk of the labor in the fish canneries along the B.C. coast, wages paid to other workers are based on what the Indians can be hired for" (2). Gender distinctions were also present with the formation of a Ladies Auxiliary of the PCFU and SPSU. The first meeting was held 1 November 1937. Ten "ladies" were present; "all mothers, daughters and wives of fishermen [were] invited to attend." (*TF* 4 Nov. 1937, 3).

The Fisherman carried a regular column entitled "The Women's Point of View" written by Donna Cogswell, who explained: "We, the wives and daughters of union members, organized ourselves into the women's auxiliaries. In helping our husbands and fathers get better working conditions we help ourselves" (14 Oct. 1937, 3). But at the same time it was being acknowledged that women were also directly involved in the fisheries. In the 6 June 1939 issue an article was entitled "Japanese Women Prepare Crabs for the King" (7), and the "Women's World" column of 10 October 1939 ran a story on Betty Louman, "First Woman to Fish Halibut" as an "Extra Man" on a crew of six Scandinavians, and the first woman member of a deep sea halibut fishing crew, on the *Norland* in the Gulf of Alaska.

All these examples demonstrate the intertwining of ethnicity, race, and gender in complicated and changing ways. While fishers were among some of the most virulent anti-Japanese racists in the province, and with attitudes against the Japanese hardening as the war progressed, Buck Suzuki became a regular contributor to *The Fisherman* (Meggs 1991, 158–9). The establishment of working-class families meant that women occupied two spheres – one in the work place and one at home – a duality which was also reflected in the union paper.

Ethnicity and race also played a part in how women were portrayed and treated, both at home and in the paid labour force.

In summary, it is important to keep in mind the durability, flexibility, and permeability of these categories. These are examples of the problematic theorized in the first two chapters whose application we can see in many different ways that yet demonstrate a certain continuum once the categories have become an accepted part of the consciousness of those who hold power, of how they come to define themselves in relation to the "other." At the same time, those categorized as "other" resist this categorization, often by force. Resistance takes on many different shapes and in this chapter I focus on resistance through unionization on the factory floor.

Gateman began Local 89's organization drive with Vancouver Island plants. Starting with the Kildonan cannery and reduction plant, by August 1941 every plant along the west coast of Vancouver Island had established locals (*TF* 1941, 5 Aug., 4; 19 Aug., 4; 2 Sept., 2; 1942, 31 Mar., 1–2).

The entry of Great Britain into World War II proved extremely lucrative for the west coast fishing industry. In 1941 the British ministry of food began to negotiate with the Canadian government for substantial portions of the fishery products (that year it secured two-thirds of B.C.'s 1941 canned salmon pack and approximately 1.6 million cases canned herring in tomato sauce [B.C. Packers 1941, n.p.]). The British also created a demand for canned herring and Vitamin A supplements produced in reduction plants. Plants and facilities were upgraded and expanded, and almost every major cannery came to have a reduction plant attached to it. Since the west coast of Vancouver Island did not have substantial sockeye runs but witnessed large herring runs throughout the war years, it re-emerged as a major fishing area in the province (just as it had during the First World War when similar conditions prevailed). Reduction plants were highly mechanized and required a relatively small but skilled staff, and they were the target of Local 89's recruitment campaign among shoreworkers. The previous chapter discussed the severe problem of pollution caused by throwing carcasses directly into the rivers. Reduction plants were an obvious solution to this problem, but they only operated when there were substantial profits to be made. At those times herring and pilchard were over-fished to the point of extinction, since in periods of high demand these were the two fisheries used to provide the raw material to the reduction industry.

When labourers other than cannery machinemen and reduction plant workers asked to join the union, Gateman tried to establish separate locals but the companies refused to negotiate with them. The unions initially tried a strategy of joint production boards. They lobbied the federal government to establish a board which would plan production and prices for the war effort and have representation from the UFFU and Local 89 and the Salmon Canners' Operating Committee (SCOC) as well as the federal government. Local 89 offered to guarantee a "no strike policy" in return for a seat on the board. The unions pointed out that "there had actually been times when through the medium of the cannery workers' organization, they had acted as a restraining force against spontaneous strikes in the plants in order to bring about amicable negotiations on wage questions with the operators concerned" (*TF* 17 Feb. 1942, 1). The federal government flatly refused, forcing Local 89 to negotiate separate agreements at every plant. At this point, the policy of Local 89 was narrowed to focus upon reduction plants and key cannerymen positions. "The union has always been willing; in fact, would prefer, to conduct negotiations only for the steady employees. While it is true our proposed agreement did cover casual labor and female labor which is seasonal and consequently very transient, this was done because a large number of the female employees who worked in canneries last year are members of the union and wish to be represented by it but we do not propose to represent the majority of this class of labour" (26 May 1942, 4).

Gateman also notes in his article that with the elimination of Japanese women from the industry there will be a shortage of "this class of labor." While he advocates that "a fair respectable basic wage for female labor should be established now," he quickly concedes that it is not a priority. "However, the union has written to the Operating Committee advising them that we are prepared to eliminate the provisions of the proposed agreement affecting female and common labor and have asked them to open negotiations for their steady cannery employees" (*TF* 26 May 1942, 4).

The "spontaneous strikes" mentioned in Gateman's report had escalated during the 1930s, as fishers expressed increasing discontent in their negotiations with canners. Shoreworkers frequently supported their demands. For example, the Deep Bay cannery (Vancouver Island) was organized in this manner. Until the 1930s, the wages and working conditions of shoreworkers had not improved substantially since the turn of the century. They worked short seasons, long hours, and were paid as little as 15 cents per hour. Author Jack Gavin (secretary of the PCFU) uses department of fisheries figures for 1933, showing there were 5362 workers employed in provincial fish canneries; 957 were white workers, 1690 were Indian, and 2715 were "of the oriental and other races" (*TF* 31 July 1937, 1). In 1897, Chinese male cannery workers received an estimated 14.8 cents per hour, or $1.48 for a ten-hour day (DMF 1902, 164). There was no minimum wage act. The only recourse to protest unbearable conditions was to stage a spontaneous strike. However, by their very nature, these strikes were sporadic and isolated events. Many were fought in order to attain parity with wages and conditions at nearby plants. But if one cannery went on strike, fish would often be taken to a neighbouring plant for processing.

The war years witnessed a deepening of the discontent that surfaced during the Depression. Salmon packs were secured at guaranteed prices, and salmon as well as herring runs were phenomenal. Canners made substantial profits, and in addition to extending production to other fisheries, like herring, they also intensified work on the salmon canning lines, hiring larger numbers of seasonal plant workers, especially women. Women began to demand better wages, and they wanted to be protected by union agreements. To cite a specific instance, in October 1941, 600 tons of herring were unloaded in one day at Imperial plant, on the Fraser River. Plant workers had to work long shifts with no overtime pay. That week, 400 men and women staged a spontaneous strike at Imperial. They voted to return to work the following shift when the company offered them an increase of 10 cents per hour, bringing women's hourly wages up to 40 cents and men's to 50 cents (*TF* 28 Oct. 1941, 1; 18 Nov., 4). Only after the dispute was settled did they join the union, a common pattern in these years. Labourers would become so dissatisfied that they would walk out or strike. Following a settlement, Local 89 would recruit the disgruntled labour forces.

Although "female and common labor" was sacrificed in order that Local 89 could concentrate on organizing permanent employees, the excluded groups continued to press for union representation. For example, in October 1942 a meeting for "all steady, cannery men" was called in Steveston on the Fraser River. So many women packed the hall that not everyone could obtain admission. All the women were turned away. "It was explained that the meeting was specifically called for steady cannery men, whereupon the women and miscellaneous workers asked when it would be possible to hold organization meetings to enrol them into the Union." Gateman explained in the article that the union's policy was based on the fact that there were two distinct sections of cannery workers; steady key cannery men employed year round in some canneries and a section of women and miscellaneous help employed only when canneries were in actual operation. In the same issue of *The Fisherman*, readers' attention was drawn to the government's adoption of a policy of "equal pay for equal work," but fishers were assured that "older unions and males generally do not need to fear that the employment of women will undermine their wage scales" (*TF* 27 Oct. 1942, 4).

Although many white fishers and plant workers held racist and sexist ideas about the rights of all labourers to decent wages and working conditions, the unions faced even more formidable opposition from canners. Employers resisted any type of union organization. However, when they began to realize that the labour movement was gaining strength and pressuring various levels of government for trade union recognition and for collective bargaining, they tried to restrict organization to the small handful of permanent employees.

The reduction plants had been selected as the opening wedge in bargaining with the SCOC since they had been the focus of the initial campaign and contained a small and stable labour force. On 28 August 1941 Gateman wrote the SCOC informing the companies of the establishment of the union and requesting a meeting (*TF* 9 Dec. 1941, 4). Meetings were held in September and in November, at which time the companies demanded formal proof that the union represented a majority of the employees for whom negotiations were in process. After submission of proof of representation the companies continued to stall by postponing meetings and generally avoiding having to deal with Local 89. In January 1942 the union applied for conciliation (*TF* 6 Jan. 1942, 4). A conciliator was appointed, and in May a reduction plant agreement was signed. By February an estimated 175 employees in eleven reduction plants were unionized (*TF* 3 Feb. 1942, 4). Local 89 then turned to negotiating for the key cannerymen and a similar pattern was repeated, again requiring a mediator. An agreement retroactive to 1 November 1942 was signed (24 Nov. 1942, 1).

The whole process was repeated the following year. An arbitration award was handed down in August. Local 89 established an 8 A.M. starting time with extra pay for hours worked beyond the guaranteed 230 per month. However, the employers refused to accept the award, and that season the cannerymen worked without a signed agreement. In 1944 the war labour relations boards came into being, easing the union's task. Cannerymen won the nine-hour day, although the 8 A.M. starting time continued to be a bone of contention. Reduction plant workers and cannerymen sought and won a guaranteed month (guaranteed to be paid for a set number of hours per month), a reduction in the number of hours constituting a guaranteed month, and overtime pay. In order to keep the guaranteed month, overtime was negotiated at straight-time rates. By 1947 guaranteed hours had been reduced to 192 per month.

Tables 7 and 8 give some indication of the number of canneries and compare numbers of shoreworkers to fishers during this period. In

Table 7
Number of Canneries in British Columbia, 1925–48

Year	Nass River	Skeena River	Rivers Inlet	Fraser River	Outlying Districts	Total
1925	3	13	11	10	28	65
1930	3	11	12	8	25	59
1935	3	9	8	10	13	43
1940	2	7	4	10	15	38
1945	1	7	1	10	10	29
1948	0	7	1	12	7	27

Source: Department of Fisheries, *Ninth Annual Report* 1925–38; *Nineteenth Annual Report* 1938–48.
Note: The year 1928 represented B.C. Packers' second bid at control of the industry through company buy-outs and subsequent plant closures.

1944 only a tiny fraction of shoreworkers had union agreements. Women were covered on a plant by plant basis, generally in the fresh fish sector where they worked as highly skilled filleters and where their numbers were small. With one exception, women never received guaranteed monthly hours. Therefore, negotiations on their behalf focused on wage hikes and the establishment of overtime pay. In 1944 the Vancouver and New Westminster plants belonging to Edmunds and Walker, Ltd signed an agreement containing an overtime clause at time-and-a-half and wage rates based on a forty-eight-hour, six-day week. This was said to be the first such agreement in the history of the Canadian fishing industry (*TF* 15 Mar. 1945, 1). The work in fresh fish plants was more evenly distributed between the men and women, especially during the war years. It was generally recognized that women were performing the same work as men on the dressing tables and in the freezer rooms but receiving lower rates of pay (27 Oct. 1942, 4; 17 May 1949, 1). Thus the principle of equal pay for work of equal value could be called upon, unlike the canning lines where sex segregation prevailed. Prince Rupert, the base for the groundfish industry, became an important centre for the organization of fresh fish workers; the concentration of fresh fish plants in a small area on the waterfront facilitated Union organization there.

In 1944 Local 89 appointed an organizer for the "1,200 Chinese cannery workers employed seasonally" (*TF* 23 May 1944, 4). Their major grievance was that contractors were attempting to have them return to the canneries at hourly rates 10 to 20 cents lower than those paid the previous year. Chan Kwan managed to recruit 200 members in four weeks, operating exclusively in the Chinese section of Vancouver (*TF* 13 June 1944, 1). At the UFAWU founding convention in 1945, Alex Gordon, who had replaced Bill Gateman as representative for the

Table 8
Number of Shoreworkers and Fishers, 1936–47

Year	*No. Employed Primary Sector (Fishing)*	*No. Employed Secondary Sector (Plants)*	*Total*
1936	11,393	6,596*	17,989
1937	11,184	5,574**	16,758
1938	10,314	6,103	16,417
1939	9,609	6,271	15,880
1940	10,444	7,449	17,893
1941	10,217	7,914	18,131
1942	12,199	6,939	19,138
1943	11,903	6,011	17,914
1944	12,426	6,150	18,576
1945	13,292	6,038	19,330
1946	13,665	6,079	19,744
1947	12,461	5,473	17,934***

Source: Report of the Provincial Fisheries Department, British Columbia 1937–48.
* Of the 6596 workers employed, 3859 were men and 2737 were women.
** Of the 5574 workers employed, 3250 were men and 2324 were women.
*** This represented a 9% decline from 1946. While employment among fishers dropped 6%, it dropped 10% in the plants.

shoreworkers, called for the complete abolishment of the "outmoded Chinese contract system" and noted that about half the thousand Chinese labourers had joined the union. The UFAWU objectives for shoreworkers that year were equal pay for equal work, a 208-hour monthly guarantee for the permanent white male employees, and decent living accommodations for Chinese workers (*TF* 15 March 1945, 1).

The Native Brotherhood, founded in 1931 and representing a wide constituency including aboriginal fishers and shoreworkers, was the first to sign an agreement covering an entire cannery crew. According to anthropologist Philip Drucker's 1958 history of the Native Brotherhood organizations founded on the northwest coast, they were formed in response to a felt need by some of the tribes to "acculturate" to white society. "Economically, the Indians had to sink or swim in an increasingly industrialized environment, in competition with whites and orientals already well adjusted to an industrial society" (Drucker 1958, 85). The Native Brotherhoods were organized to imitate "white models" as a response to the competing claims from the white population to resources necessary to maintain the aboriginal coastal economies. These organizations were apparently also heavily influenced by the missionaries, especially Duncan's work at Metlakatla. Finally, the model of the trade union was also investigated and partially adopted.

Alfred Adams, from Masset on the Queen Charlotte Islands, founded the Native Brotherhood in British Columbia after he attended two conventions of the Alaska Native Brotherhood. He investigated the possibility of an "intertribal organization" among Port Simpson fishers. In December 1931 meetings were also held with representatives from the villages of Hartley Bay, Kitkatla, Port Essington, and Metlakatla. The organization grew, and in 1942 it merged with the Pacific Coast Native Fishermen's Association. The merger changed the character of the organization, involving it more directly in bargaining on behalf of a large number of aboriginal commercial fishers. Drucker notes, however, that the Native Brotherhood was a "conservative" force in comparison to the "aggressive white union; hence the industry gave the Indian organization recognition quite willingly" (Drucker 1958, 109). Each village with a Native Brotherhood branch also had a Native Sisterhood "which is simply a women's auxiliary of the local branch" (116). Attempts were made in the late 1930s to establish an overall Native Sisterhood modelled on the one in Alaska but there was resistance from villages like Port Simpson.

> There are still a good many Indian women who work in the canneries during the season, either in connection with the packing operation or in the net lofts. Instead of the standard $2 per year Sisterhood fee, these women pay $3 a year to the Vancouver office. In return, their wages and working conditions are supposed to be provided for through negotiations with the canning companies by the business agent and Executive Committee. Actually in recent years it appears that the canning companies, simply through courtesy, make the same agreements for cannery workers with the Brotherhood that are made with the white union ... Wage scales set for net women are arrived only with the Brotherhood since it is said there are no white women who do this work. (Drucker 1958, 118)

In 1943 the Native Brotherhood signed with five companies for plants that employed aboriginal labourers. But when the UFAWU began to negotiate on behalf of all cannery employees, the Native Brotherhood tended to adopt the same agreements for its membership, and the UFAWU agreed that aboriginal labourers could be joint members in the two organizations.

DECIMATION OF THE CHINESE LABOUR FORCE

While the Native Brotherhood spearheaded unionization among aboriginal cannery labourers, in May 1944 Local 89 began a serious attempt to organize an estimated 1200 Chinese cannery workers. Recruiting in Vancouver's China town alone, its Chinese organizer was able to organize approximately 200 workers (TF 13 June 1944, 1). Local 89 continued to appoint Chinese organizers until 1949, when very few Chinese remained in the industry. Within a very few years an enormous number appear to have disappeared. Employers (J.H. Todd, 1947–50), when faced with the abolishment of the Chinese contract system, eliminated their Chinese employees along with the contractors.

For example, in his report for 1946, the organizer estimated that between 1000 and 1500 Chinese were employed in the fish canneries, their numbers exceeding those in any other basic industry with the possible exception of forestry (TF 13 Sept. 1946, 3). He noted that the organizational drive that year in northern canneries had been unsuccessful. The Chinese workers expressed concern that agreements covering themselves and women had not been signed. They also questioned the union's strategy of organizing on a sectional rather than on a plant-wide basis (UFAWU 1946, vol. 190, p. 2 of letter dated 19 Aug.). In the 13 September 1946 issue of *The Fisherman*, organizer Alfred Quan reported on the motives to unionize the Chinese which included protecting and raising their standards of living and "removing the danger of Oriental cheap labor undermining our standard of living" (3). But he listed four adverse factors. First, the seasonal nature of the work with three to four months' employment led to economic insecurity and instability and unrest among the Chinese labourers. Second, the contract system meant that workers depended on the contractor for their means of livelihood during the months when the canneries were closed. Third, Quan listed the "peculiar temperament" of the Chinese worker who tends to prefer "sentimentalism and nepotism" (in reference to the tradition of hiring relatives) and thus shies away from labour unions. And the last factor is the lack of sincere attempts by white workers in the past to help them better their standard of living. The specific difficulties Quan experienced on his northern tour included inadequate transportation facilities, the absence of an agreement to cover Chinese workers, a lack of interest on the part of white workers regarding the welfare of the Chinese, and the influence of the Chinese boss. He did not judge his two predecessors to have been particularly effective (3).

In 1947 (4 July) *The Fisherman* listed the cannery machine men's supplement with the top monthly rate of $225 for linemen and chief engineers (5). Casual Labor employed on a monthly basis received $150. There was also a "Chinese Supplement" (8). While general labor was also paid $150 per month, the highest monthly wage went to butchers, Iron Chink feeders, and unloaders, who received $163.20.

Far fewer classifications were listed in this supplement than for the cannery machine men. Meanwhile, the classification "Women Cannery" contained only hourly and piece rates for women. Only three hourly rates were listed: 55 cents for "Beginners (first 48 hours)," with 82½ cents overtime; 65 cents for "Slimers, Washers and Lineworkers" (97½ cents overtime); and 70 cents for "Labelling Machine Operators" ($1.05 overtime) (4 July 1947, 5).

In 1948 the organizer estimated that Chinese labour had been reduced between 30 and 35 percent. Many labourers had apparently returned to China. In addition, the cannery labour force was well advanced in terms of age (since immigration laws prevented further migration into the province after 1923). However, Quan attributed the main reason for the decline to an avowed policy by the companies that they were being more selective in hiring Chinese labourers now that they were included in union agreements (UFAWU 1948, vol. 190, p. 2 of report).

The UFAWU opposed the Chinese contract system. In seeking plant certification, the union would have been forced to negotiate with the contractors, since they recruited and paid the crews. It could only represent employees and bargain on their behalf with the SCOC once they appeared on company payrolls (UFAWU 1947, vol. 190, letter dated 18 March). But provincial labour legislation was already being enacted that would force employers to list their employees by name and pay them directly. The SCOC, seeing the writing on the wall, advised companies in 1947 to accept UFAWU certification. "This Agreement, which is now ready for signature, is an entirely new departure from the practice in effect for so many years respecting the employment of Chinese workers. It is our view that these workers should be considered as direct employees of the company, and, in fact, this is the ruling of the Provincial Department of Labour" (J.H. Todd 1947, Box 21, letter dated 24 June). Forced to recognize Chinese labourers as employees with a right to be unionized, many companies stopped employing them altogether (J.H. Todd 1948, Box 21, letter dated 17 June).

Despite union certification, the Chinese contract system was not completely abolished. Rather it was modified to meet provincial standards. Contractors continued to recruit seasonal labour and to provide direct supervision. The foremen, or "China bosses," were not placed on the cannery payroll but were paid directly through the company's head office (J.H. Todd 1947, Box 21, letter no. 21 dated 8 July). The consequence for Chinese workers was that they were now covered by two contradictory systems. Under the old Chinese contracts, although they were paid substandard wages, they were hired and paid for the entire season. Many labourers also depended on contractors to find them work during the rest of the year. Being covered by union agreements did not sever this relationship. However, under UFAWU agreements, they lost their seasonal guarantees, although the union managed to secure monthly guarantees (UFAWU 1948, vol. 190, notes dated 15 July to 20 Aug.). And when employees were listed individually on payrolls, employers began to pick and choose, discarding older workers.

As the various groups of seasonal labourers became certified, special supplements to the master agreement were written to cover them. While the master agreement elaborated general working conditions, the supplements listed wages for specific categories of labourers. In 1946 women working in canneries were included for the first time. The companies made further distinctions and kept wages and categories for Native and white women separate. While white women were paid by the hour, Native women received piece rates in jobs like hand filling cans of fish (J.H. Todd 1947–49, Box 21 contains cannery agreements for Inverness, Klemtu and Empire canneries). In 1947 agreements covering Chinese cannery labour were first signed and a separate supplement was drawn up for them. Two years later they were included in the Male Cannery Workers' Supplement, which was divided into four classifications. While the top wage in Groups I and II was $245 per month, Chinese labour was paid under Group IV with a top wage of $183 per month (file labelled "1949 questionnaires"). A summary of data covering canneries owned by J.H. Todd & Sons (Klemtu on Finlayson Channel, Inverness on the Skeena and Empire on Vancouver Island) in response to a labour relations questionnaire compiled by the SCOC for 1947 and 1948 appears in the appendix. Todd also owned Richmond cannery located on the Fraser and Beaver Cannery on Rivers Inlet, but there were no canning operations there during these two years. The nineteenth-century structures are clear in the job classifications according to race and gender, by the pay scales and guarantees of employment (piece rates, hourly rates and monthly guarantees, depending on the race and gender of the group covered), and even in the segregated housing (with attendent services) available to each group.

By 1949 virtually the entire provincial salmon cannery labour force was unionized. However, the divisions salmon canners had developed allowing them to pay individual groups of labourers the lowest wages possible for that particular category of labour were incorporated into these first union agreements. Since fishers had initiated the move toward an industrial trade union, their interests predominated. They wanted provincially unionized plants, and the best way to achieve this goal was to recruit from the top, from among those labour categories judged to be indispensable to the companies. Without the war and the ensuing intervention of the provincial and federal governments in passing labour legislation and acting as arbitrators, it is unlikely that Local 89, and its successor the UFAWU, would have had any success in negotiating for shoreworkers. In the process, however, the UFAWU mirrored the sexist and racist structures the companies had established. As Jack Nichol, who replaced Alex Gordon as shoreworker organizer and in 1977 became the first shoreworker to be president of the UFAWU, remarked: "UFAWU cannery agreements were one of the most notorious examples of discrimination against women in employment. Two supplementary wage agreements, one colored blue and the other pink, set out wage conditions for men and women" (*TF* 17 July 1973, 3), The next task for union shoreworker organizers was to eradicate racial and gender discrimination contained in their own union agreements. They began with racial discrimination against aboriginal workers.

COMPANY RETRENCHMENT AND LABOUR PASSIVITY: THE 1950S

During the 1950s, the interests of fishers and of the small core of permanent employees prevailed within the union, illustrating the enormous differences between the conditions of white male and seasonally employed labourers. Both the UFAWU and the fishing companies experienced a number of setbacks following the end of the Second World War.

Once the war ended, markets became problematic. In 1948 the canning of herring came to an end. It had lengthened the employment of seasonal workers since it preceded salmon canning. During the 1930s, companies had begun a campaign to close remote outlying canneries, a trend they halted during the years of wartime prosperity but resumed after the war. Closure of canneries was accompanied by a reduction in labour forces. For example, whereas in 1944 a total of 6150 shoreworkers were employed, the number fell to 3947 by 1952 and subsequently continued to fall (*TF* 8 June 1954, 1). While in 1944 there were a total of 31 canneries with 100 canning lines, by 1957 these were reduced to 17 plants with 58 lines (*TF* 14 March 1958, 3). Operations were consolidated and centralized in large plants located near or in Vancouver and Prince Rupert. It is evident from the oral histories summarized in chapter 4 that the closure of canneries in the north (along the Nass and Skeena rivers), along the central coast (including Rivers Inlet), and on Vancouver Island cut off employment opportunities for a large number of aboriginal villages that had come to depend on shore work to sustain their economies (Drucker 1958,

125). The UFAWU began to mobilize seasonal labourers around this issue.

While the majority of aboriginal fishers, especially those in the big boat fleet, were men, many aboriginal women became shoreworkers, and in becoming proletarianized through wage labour, they developed a class consciousness that created cleavages within many households. Overall, aboriginal women were stauncher UFAWU supporters than their male kin who more readily espoused the causes of organizations like the Native Brotherhood of British Columbia (NBBC). Referring to interviews conducted during the course of the Fish and Ships Project in the early 1980s, Pinkerton (1987b, 261) observed: "Virtually all Indian shoreworkers are in the United Fishermen and Allied Workers Union and appear, from our interviews, to maintain a more militant and consistent trade-union stance than do Indian fishers. As well, those interviewed said they had more in common with non-Indian shoreworkers than with Indian fishers in the Native Brotherhood. It was not uncommon to hear an Indian fisher express negative or neutral attitudes about the UFAWU and be interrupted by his shoreworker wife, who wished to express her support for the union." Marchak documented what was at the time the latest in a series of disputes between the NBBC and the UFAWU. The executive director was quoted in Vancouver newspapers as claiming that the "union had sold the fishermen out to get a raise for shoreworkers" (Marchak 1987, 236–40).

While the UFAWU and the NBBC negotiated jointly for collective agreements, the union focused on the proletarian interests of shoreworkers and fishers, while the mandate of the NBBC was considerably wider. Aboriginal fishers had interests tied to their aboriginal claims and status. In addition, large boat owners who hired wage labourers were also eligible for membership in the Native Brotherhood but could not join the UFAWU. These differences led to severe disagreements over the years. For example, the salmon strike in 1952 ended after four days to prevent growing disunity and divisions among fishers. The NBBC opposed strike action and indicated it would sign agreements with the companies for its members.

Drucker (1958, 131) referred to a series of negotiations between the UFFU and the NBBC which attempted to include the latter in collective bargaining. In the end, the two organizations agreed to negotiate jointly with canners but to sign separate agreements. However, in 1945 when the NBBC received its charter, "union leaders lodged an official protest to the Trade and Labor Congress of Canada against an organization chartered under the Societies Act rather than as a labor union being permitted to sign labor contracts" (131). When the NBBC protested, the union denied that the Native Brotherhood had been its target. Drucker (149–50) attributed the conservative attitude of the NBBC to a feeling of loyalty to canners "who for many years have been the only source of credit and thus have provided the only means through which the individual Indian could better himself financially." Another important consideration was the fact that aboriginal fishers were far more dependent on fishing than white fishers, who in the event of a strike that lasted the entire season could often find other employment.

Conflicts with aboriginal fishers were not the only source of division between groups of fishers and the union. Gear type and amount of capital investment in boats and equipment continued to divide fishers. In addition, the legal status of the union in representing vessel owners was challenged several times by recourse to anti-combines legislation. Fishers were said to be "co-adventurers" with fishing companies rather than wage labourers. The UFAWU withstood such charges by pointing to its representation of shoreworkers and tendermen.

The U.S. fishing industry on the northwest coast had unionized much earlier, but stronger anti-combines legislation decimated those unions after the war (Meggs 1991, 167–74). By 1952 "the UFAWU was standing alone as the single functioning industrial union anywhere in the fishing industries of North America" (*The Fisherman* 2 Feb. 1979, 10). The constant struggle by employers to reinstitute cheap wage labour was made evident in an article also contained in this issue of the union paper.

In Alaska, companies were fighting ILWU organizational drives among shoreworkers by using "floating factory ships," thus disregarding safety, minimum wage laws, and the forty-hour week. Labour contractors operating out of Washington state and California were recruiting among recent immigrants from the Phillipines, Vietnam, Korea, and Latin America, people who spoke very little English. These people signed a personal service contract and were issued a one-way ticket to Dutch Harbor, an isolated community 1200 kilometres southeast of Anchorage in the Aleutian Islands. Their return fare would be paid only if they completed six months' work, otherwise the cost of the plane ticket was deducted from their pay. They also had to pay $100 damage deposit for their accommodation. They were paid $3.50 an hour and worked twelve hours a day, six days a week. Although the minimum wage was $3.40, some contractors negotiated for labourers to work for $1000 per month (for an eighty-four-hour week). Anyone who became sick or was injured and thus unable to work had to pay $20 a day for room and board. Serious industrial accidents due to the long hours were reported. "By floating their plants just feet from the Alaska shore, the owners evade federal occupational health and safety regulations ... as well as legal standards for accommodation" (*TF* 2 Feb. 1979, 3). Some workers were forced to sleep on the floor while women were at risk of being raped by foremen and other workers and were warned to keep a club by their beds. These conditions were being reported as occurring in the late 1970s.

That same issue of *The Fisherman* contains a short history of the American and Canadian governments' legal challenges of fishing unions. In British Columbia, the attacks began in 1952 when two gillnetters were unsuccessful in bringing a charge before the B.C. Supreme Court that the UFAWU was not a trade union. The challenge was taken up again in 1954, following a strike by salmon tendermen, when four gillnetters charged that the UFAWU was an unlawful combine. In 1956, union offices were raided and many documents seized in yet another investigation (10–11).

These events were partially connected to the political affiliation of the UFAWU leadership. The Trades and Labour Congress, as part of a wider movement to disbar all unions affiliated with the Communist Party, suspended the union from 1953 to 1973 because of its communist leadership and orientation. In November 1950 two executive members of the UFAWU were barred from attending the Victoria Trades and Labour Congress because of their political affiliation. Shoreworkers, especially aboriginal labourers, struggled with this issue, which was exacerbated by the fact that many aboriginal people had become Christians and had difficulty accepting the affiliation of union leaders with the Communist Party.

All these problems threatened to tear the union apart. In fact, splits among fishers were aggravated over the next few decades while shoreworkers began to articulate their own interests and to organize collectively around them. However, these interests had to be formulated not simply for and by the small core of permanent employees but also by the large group of seasonally employed labourers.

The most militant shoreworkers were reduction plant workers, cannery machinemen, monthly classified netmen, and fresh fish and cold storage workers. Most of them were men, and many were also commercial fishers. They worked closely together and were permanently employed, often alternating employment between fishing and work inside the plants. Their interests were linked to those of fishers, and thus they were receptive to the fishers' organizing drives. These men were also craft workers who could point to disparities between their wages and working conditions and those of their counterparts in other organized provincial industries.

During the 1950s and 1960s, these groups realized significant gains. Their guaranteed monthly hours slowly decreased to the equivalent of a forty-hour week. They fought for and won multiple overtime rates. They sought parity with their fellow tradesmen in other B.C. resource industries. Although reluctant, the Fisheries Association (the SCOC changed its name in the early 1950s) was willing to negotiate better conditions for them. By demanding percentage increases in pay, this militant sector widened the gap between its wages and those of the general labour and female categories. Ultimately, the Fisheries Association was willing to negotiate because this sector represented such a small number of employees. For example, in 1954 the cost of requests by the shore sector (a forty-hour week with no reduction in take-home pay) was estimated to represent one-fifth of a cent per half-pound can of salmon (*TF* 13 July 1954, 1). This figure was calculated on the basis of 300 monthly paid men receiving seven months employment and 12 percent overtime; 600 hourly paid men employed for 3.5 months and receiving 10 percent overtime; and 1400 women employed for 2.5 months and receiving 10 percent overtime. The 300 men in the monthly paid category were the militant sector, but according to these figures, they represented only 13 percent of the labour force.

Union organizers found the Fisheries Association adamant in its refusal to grant any wage increases to seasonally employed women and to casual labourers who together formed the bulk of the plant labour force. These two groups went without any wage increases between 1951 and 1953, and faced very different conditions of employment. The government's policy of securing guaranteed packs ended in 1947, and until the late 1950s companies had a difficult time re-establishing domestic markets and finding new outlets for their commodities. They thus found themselves with a surplus labour supply and could resist wage demands. In addition, mechanization occurred throughout the 1950s, a process that allowed canners to close outlying canneries and to consolidate their operations in a few large plants. These plants became multi-product and multi-processing operations which combined diverse processing techniques and product lines in locations accessible to transport systems for the final shipping of the finished commodities. Despite cutbacks in their numbers, perhaps because they feared employers would fire militant seasonal employees, male casual labourers and cannery women refused to take strike action.

THE CAMPAIGN TO END RACIAL DISCRIMINATION

Despite all these divisive interests and conflicts, the UFAWU was founded on democratic principles. In 1945, for the first time since the UFFU had been created, women and Chinese labourers were represented among the delegates to the annual convention (TF 6 Mar. 1945, 1). Alex Gordon replaced Bill Gateman as shoreworker organizer in 1944. Gordon began to attempt to organize and press forward the demands of seasonal labourers. Beginning in 1950 the union launched a major campaign to end racial discrimination. The editorial in the 21 February 1950 issue was entitled "Equality for Natives." "Not only should our Indian people be given the franchise, but they should also have the advantage of such social security measures as the Old Age Pension ... We wish to point out, however, that discrimination on the basis of racial background is a common method used to divide workers" (21 Feb. 1950, 4). "We must not forget the lessons of the past when the fishermen and allied workers were divided along racial, gear and craft lines and the operators were thus able to play one group against the other to the detriment of all" (6 Jan. 1953, 2). Secretary-Treasurer Homer Stevens was denouncing those "persons who would like to establish splinter groups and in fact to split the UFAWU wide open ... They try to pit Native fishermen against non-Native fishermen, gillnetters against seiners, tendermen and shoreworkers against all fishermen" (1). These were the same people Stevens identified as targetting the "reds who run the Union."

While shoreworkers were reluctant to fight for better wages, they (especially aboriginal women) were prepared to fight on other issues, including better company housing and payment of transportation costs to and from the plants. At most plants, housing for aboriginal people reflected the racial discrimination upon which the industry was built. For example, at Namu, on the central coast: "The matter of two Native women who came to the plant early and were placed in the regular bunkhouse and were then compelled to move into the separate bunkhouse for Native women later on was raised as an instance of what we consider to be discrimination" (UFAWU 1956, vol. 52, minutes of meeting held 28 Aug.). Seasonal labourers learned to participate in union meetings by raising individual concerns of particular importance to them. These then provided a forum both for discussing the problems pertinent to specific groups of people and in which workers could develop leadership skills. In 1954 a symbolic act took place at Namu, when a meeting of 154 NBBC and UFAWU cannery workers voted unanimously to remove the signs on two adjoining women's rest rooms, one reading "Natives" and the other "Whites." The increasing militancy of aboriginal women was reflected when they changed their status within the Native Brotherhood from a Women's Auxiliary to the Native Sisterhood. The Native Sisterhood was being absorbed back into the NBBC. At the 1955 UFAWU convention, it was noted that native women had stood shoulder to shoulder with their men and had fought "the good fight for better working conditions in the fishing industry." Ellen Neel, a speaker at the delegation representing the Native Sisterhood, affirmed: "We hope and trust that the time is not too far distant when every sister will have an equal voice with every brother" (*TF* 5 April 1955, 3). The UFAWU planned eventually to have women on the negotiating committees dealing with plant working conditions. "Who but the workers themselves have a better knowledge of these working conditions than those people actually engaged in the work itself?" (3).

It was not until 1956 that the agreement embodied the principle of equal treatment of aboriginal workers. Until that time, "guarantees, paid transportation, seniority, and board conditions do not apply to Native women at B.C. fish canneries" (*TF* 17 April 1956, 7). The NBBC was working with the UFAWU on this issue, which also included the need for nurses and personnel to care for the children of working mothers at out-of-town plants. While Chinese and other male labourers were transported to outlying canneries at company expense, women were assumed to be travelling with male relatives (for example, Japanese women accompanying their husbands on fishing boats) or to be living near the sites, and were held responsible for their own means and costs of transportation. In 1949, a small concession was won when women sent to northern plants were given a monthly guarantee: "With this amount to be payable to all women cannery workers at such plants regardless of the place of hiring" (7 Mar. 1950, 8). Gordon had less success, however, in abolishing the piece rate system. For example, in 1952, aboriginal women were required to fill 346 cans per hour to achieve the minimum hourly wage of 98 cents (22 Jan. 1952, 6). Gordon wanted the rate brought down to the equivalent of 250 quarter- or half-pound cans per hour, the piece rates earned by "slimers, washers and line workers." Although the union was able to institute beginners' rates for this category of labour, piece rates were not abolished until the 1960s when the adoption of quarter-pound filling machines rendered the hand-filling of salmon tins obsolete.

While aboriginal women were particularly hard hit by the closures of northern and outlying canneries, women workers in general suffered from company retrenchment policies. Although the number of labourers declined, production levels remained constant. High-speed machines were adopted on the canning lines, existing machinery was speeded up, and new machines were introduced that allowed the companies to hire fewer labourers, especially in the warehouses, can lofts, and on the filling lines. In the period 1944 to 1953 an estimated 50 percent fewer women were hired. For example, the installation of high-speed machines in one cannery led to a reduction in the crew from sixty to forty-eight workers, while the same volume of fish was processed in less time (*TF* 1954, 20 April, 1; 13 July, 6).

Speed-ups affected not only the labour of women on the cannery lines. In 1953 filleters working in a Prince Rupert plant were told that they must meet production standards and were offered a 6 cent per hour bonus if they could surpass the standard required of an experienced filleter (forty-five pounds of sole, skinned both sides, per hour). The company in question was also contemplating introducing piece rates on an incentive bonus plan for picking shrimp or crab where such operations were sizeable (*TF* 19 May 1953, 1). Women had to work more intensively in those areas where their work paced them to the speed of a machine, as well as in those jobs which remained manual operations. Gordon used this to fight for better wages. "Any comment with respect to the type of work done and general conditions under which the women work is unnecessary for anyone familiar with the industry. Certainly, the work is more difficult, wet, and dirty, and with just as much skill required, as work done in the sugar refineries, paper box and bag factories, and meat packing plants." He was negotiating for a general wage increase of 10 cents per hour for women. "In what other industry are women required to display the same skill and strength for the kind of wages paid to experienced filleters and for the type of sporadic employment offered?" (3 March 1953, 1, 6).

Nevertheless female plant workers did not fare well on the wage front in the 1950s compared to the gains made by other categories of labour and by fishers. Employers refused to recognize their right to decent wages, and women themselves found it difficult to press their demands for higher pay. It is important to place these difficulties within the theoretical context discussed in the first two chapters. In other words, women (especially aboriginal women) were not simply or only struggling for economic gains. These gains were embedded in a whole structure that devalued their labour and who they were as people. Thus their fight for economic equality was at the same time a struggle for recognition as persons of equal rank to the men (employers, fishers, and permanent plant employees) who held power in the industry. At the same time, women had to fight against their own socialized internalization of inferiority.

While fishers went out on strike year after year, shoreworkers adopted a conciliatory stance. The most militant were the ones who realized gains. Although Gordon pressed for a 10 cent per hour increase for women in 1953, in the final agreement they received no increases at all, unlike the skilled cannerymen, netmen, and refrigeration engineers (*TF* 1953, 3 Mar., 1, 6; 7 July, 1) whose gains the fishers jealously regarded. "Several fishermen have asked what the proposed increases so far granted to shoreworkers will mean in increased costs to the fishing companies" (16 June 1953, 2). In a tightened economic climate each group sought to protect and promote its own interests first. Women did not push for their demands; as a result, while plans to improve their wages and conditions were formulated, they were often the first to be dropped at the bargaining table.

Beginning in 1952 Gordon held annual wage conferences early in the year. Male and female cannery workers, networkers, fresh fish, and cold storage workers, those in the reduction plants, steam and refrigeration engineers, watchmen and saltery workers attended on a representational basis. At the first of these conferences, some groups expressed more dissatisfaction than others with existing conditions. The most vocal were the craftsmen. Gordon noted that women needed wage increases because the cost of living had escalated and speed-ups had been instituted in the plants. "Among this group, however, it was recognized by the wage conference that dissatisfaction with the rate during the 1951 season was not strongly expressed" (*TF* 22 Jan. 1952, 1). By holding meetings on a representational basis, it was hoped that those sectors making the least gains would begin to fight for their rights. Otherwise, they tended to be drowned out by those who were more vocal at meetings, usually those who had made the most gains.

Two sets of arguments were used in negotiations. Comparisons were drawn between craft workers in the fishing industry and those employed in, for example, sawmills, logging camps, smelters, pulp and sulfite plants, shipyards, and the building trades. Employers were willing to grant percentage increases with the highest rates going to the trade classifications. Beginning in 1951, however, the union began to insist that all classifications receive the same increases (*TF* 20 Feb. 1951, 1). Company representatives opposed this move, arguing that it was not logical and that the union should request graduated increases with the highest rates receiving the best percentage raises (20 April 1954, 1).

By fighting for equal across-the-board raises in pay, Gordon sought to narrow the large gap between monthly paid craftsmen and the seasonally employed, hourly paid labour force. He also advanced a second set of arguments, insisting that wages paid to women should be comparable to those received by women in such *organized* industries as bakeries, pulp sulfite plants, B.C. Sugar, American Can Company, the distilleries, plywood plants, and supermarkets. As late as 1958 women's wages lagged far behind those in these industries. Their wages also lagged behind the much older unionized fish plants in Alaska and Washington. But the companies insisted instead on comparing wages with those of fruit and vegetable cannery workers in the Vancouver and Okanagan areas, certain food packaging plants, laundries, selected classifications in meat packaging plants, certain gold mines, fishworker rates in Nova Scotia, and rates paid to workers in Japan and other fish-producing countries (*TF* 1958, 14 Mar., 3; 20 June, 3). In other words, employers insisted on comparing women's wages to those received by women working in unorganized industries.

Gordon's efforts for real gains for seasonally employed women also serve as a basis to criticize a theory that has become very popular in explaining differential wages in this type of industry: split labour market theory (Bonacich 1976, 1979, 1980; Edwards 1979; Piore 1979). Bonacich argues that there are in fact two labour markets. The primary sector includes unionized craft white male workers, while the secondary sector incorporates non-unionized, seasonal labour forces recruited among African-Americans in the United States. I have deliberately chosen not to use this theory as an explanation (B.C. fishers and shoreworkers would be obvious examples of both sectors) because it is descriptive (it does not account for how these divisions arose in the first place) and because the efforts of the UFAWU during the 1950s and beyond clearly attempted to overcome these distinctions, which it had inherited from salmon canners in the nineteenth century. I believe that a theoretical conceptualization of cheap wage labour, with all its complexities in incorporating patriarchy, class relations, colonialism, and imperialism, better reflects the "problematic" at work in industries such as this one.

INCREASING MILITANCY: THE GENERAL STRIKES OF 1959 AND 1963

In the two-year agreement of 1954, the union finally won the same rates of increase across the board for all shoreworkers, and, effective the following year, the forty-hour week (*TF* 17 Aug. 1954, 1). "To our knowledge, no other definitely seasonal groups either in Canada or the U.S. have established the condition" (3 May 1955, 2). In 1956, equal conditions for all racial groups and equality between male and female labourers were cited as key objectives at the initial negotiating meeting for cannery, net, and reduction plant workers. Included were improved accommodation facilities for aboriginal workers and provision of nursery and playground facilities for children. The three groups decided to settle their agreements together that year, narrowly averting a strike. "And just as shoreworkers have backed fishermen to the limit in their many struggles against the operators, so will the fishing fleet stand solidly by their fellow members of the Union in the shore section of the industry" (*TF* 10 July 1956, 4). The final two-year agreement contained another across-the-board increase for all shoreworkers. In 1957 a welfare plan was instituted, and equal treatment of aboriginal workers was written into the collective agreement (1956, 14 Feb., 1, 6; 10 July, 4; 17 July, 1, 6). Shoreworkers did not negotiate in 1957, but fishers were involved in five strikes that year. The companies tried to resolve the net fishers' strike by creating friction between fishers and shoreworkers.

When the net fishers went on strike in 1957, the companies closed the canneries. They also laid off monthly employed networkers and cannerymen. In the meantime, fishers had to work on their gear, and the companies hoped this would create internal conflict. But fishers and networkers arrived at a compromise whereby fishers could work on their own gear while all company gear was declared untouchable by anyone but networkers (*TF* 2 July 1957, 1, 4). When the strike was settled, the 150 monthly employed shoreworkers remained locked out of the plants. A ballot conducted among this group resulted in a 64 percent vote in favour of a complete walkout, less than the required two-thirds majority. The issue went to conciliation with the result that shoreworkers lost their pay for that period. A split board award ruled that monthly workers were subject to sporadic layoff (7 Feb. 1958, 1).

In 1958 the union was forced to accept percentage increases, although it insisted on a one-year contract. Following a decade with improved conditions but few wage increases, by now the membership was becoming increasingly militant. Thus in 1959 union policy was to sign no agreements unless all agreements covering fishers, tendermen, and shoreworkers were signed at the same time (*TF* 10 April 1959, 1). From 29 July to 7 August that year there was an industry-wide strike in which all groups actively participated. This was the first general strike in the provincial fishing industry. It was short lived, ending when certain groups began to weaken, the first being the shore section in Prince Rupert. In support of the strike, crews at Tulloch-Western, an "independent" company which bargained separately from the Fisheries Association, canned salmon using fish caught during the strike and placed in storage when the government prevented fishers from selling it to the public from the docks at Vancouver Harbour. This canned salmon bore a "strike relief" label, and proceeds from its sale went toward strike funds. It proved to be a very popular and effective strategy to publicize strikes in the fishing industry (21 August 1959, 1).

By the end of the decade, shoreworkers emerged as a much more cohesive group. Their ranks had been decimated by plant closures, companies consolidating and centralizing their operations, mechanization, and the speed-up of work. However, those plant workers who were still employed had won important non-monetary benefits. Especially significant was the institution of seniority rights for seasonally employed workers which guaranteed that they would be rehired each season. Racial discrimination was identified, brought into the open, and fought. In the decades that followed, equality in the wages paid to men and women continued to be an important issue. In the 1950s a number of women emerged as strong leaders. In 1954 Mickey Beagle, elected second vice-president by acclamation, became the first woman to be elected to a UFAWU office (6 April 1954, 1). In 1959 she assumed a full-time union position as a general organizer. In 1956 Verna Parkins was elected president of the Prince Rupert Shoreworkers local, and this was believed to be the first time that a woman had headed a local (31 Jan. 1956, 1).

Although the shore section was the first to weaken during the 1959 general strike, the strike itself marked a turning point. Fishers, tendermen, and shoreworkers began to insist that all agreements be negotiated concurrently, thus preventing the companies from playing one group against the others, for example, by claiming that fishers had negotiated fish prices too highly and therefore employers could not afford to give increases to tendermen or shoreworkers. As all three groups began to work collectively in negotiations, shoreworkers began to press their interests within the union, and the strength of this sector began to grow over the next decades.

The companies continued to try to divide union members against one another. As shoreworkers became more militant, employers made a point of hiring large numbers of casual employees, often students, when a strike vote was being conducted. For example, in 1962, 53.1 percent of the plant workers voted to strike. But 74.4 percent of fresh fish and cold storage workers favoured strike action. The votes among cannery workers were inflated because an extra 897 workers were hired (compared to the numbers hired in 1959) to cope with the large run of pink salmon. Gordon insisted these people had little if any previous experience. By holding a vote at this time, only 47.8 percent of the combined cannery, net, and reduction classifications voted to strike (*TF* 30 July 1962, 1). But in the following year, 1963, the industry experienced a second general strike. While shoreworkers voted to accept a conciliation award, the Fisheries Association turned it down. Once again, the companies hired a large number of workers for one or two days during the voting period, but this time they could not prevent a strike vote. They responded by threatening to close down for the entire season. While shoreworkers and tendermen settled, the net fleet sailed without an agreement. The outstanding issues were referred to an arbitrator, which resulted in the appointment in 1964 of a federal-provincial committee to investigate the entire area of negotiations over fish prices. It recommended compulsory arbitration.

The threat of compulsory arbitration dampened the net fishers' desire to strike the following year. The numerous strikes of the previous few years had aggravated tensions among fishers, and this was reflected in poor attendance at union meetings. Fishers would not have realized any gains in 1964 if shoreworkers and tendermen had not voted to cancel their newly negotiated agreements in order to support the demands of the net fleet. A crisis was brewing among fishers, and it erupted two years later.

Shoreworkers had not been the only ones affected by mechanization. The introduction of brine systems on fish packers resulted in the elimination of tendermen from herring packers. The tenderboat sector was intermediate between fishers and plant workers, emerging when fishboats roamed ever more widely and required vessels to carry their loads back to the plants and to unload them from the packers onto the docks. The smallest of the three sectors, mechanization steadily eroded its membership (Stacey 1982). More jobs were eliminated with the introduction of air pumps to mechanically unload fish from packers and fish boats. Until then, this had been the most manually intensive job for tendermen. The labour force was decimated even further with the introduction of refrigerated packers and when companies increased their use of contracted tug boats and refrigerated trucks to transport catches. By the 1970s tendermen were fighting for the very survival of their trade.

These technological changes also affected the fishing fleet. New refrigeration systems as well as improved engines and the use of electronic equipment encouraged the emergence of a big boat fleet. Boats could now combine gears, entailing considerable capital investment. By combining gears (for example, gillnets with seine nets) fishers could also combine fisheries such as salmon and herring. The development of a fresh/frozen market (in its turn facilitated by the adoption of improved refrigeration techniques) stimulated the growth of the groundfish industry, historically associated with the emergence of large trawlers on both seaboards.

The economic slump of the 1950s lifted in the following decade. While some fishers plowed their profits into their vessels and became small capitalists, canners looked for new fisheries to exploit. B.C. Packers began to expand its groundfish processing operations. In 1962 the Weston food conglomerate took control, and B.C. Packers expanded its groundfish operations further, from the Pacific to the Atlantic.

The UFAWU viewed both these developments – the increasing importance of the big boat fleet over which it held no jurisdiction (any vessel owner employing more than two crew members could not be a union member) and the expansion of B.C. Packers into the groundfish sector – with great alarm. In January 1967 the union began a disastrous attempt to organize trawl fleets in the Maritime provinces, resulting in a seven-month strike in 1970 against Booth and Acadia Fisheries. The companies threatened to close the plants completely, declaring them to be uneconomical to operate. Fishers lost the entire season, workers lost their jobs, and the UFAWU was driven from the east coast.

During the same time the union was trying to gain admission to the Canadian Labour Congress (CLC). The CLC stipulated that it would only accept the UFAWU if it were to affiliate with an existing charter member. Mergers were contemplated not only in order to gain admission to the CLC but also in the belief that the only way to challenge the giant food conglomerates was through a strong national union of food workers, just as in the provincial fishing industry strength lay in an industrial province-wide trade union. But negotiations with other unions collapsed over the question of how much autonomy and power the UFAWU would retain within the proposed alliances. Nevertheless, in 1973 the CLC agreed to admit the UFAWU on condition that it relinquish its attempts to organize on the Atlantic coast.

The UFAWU also tried to organize groundfish operations on the Pacific coast. The two gear types involved, trawl and longline, were already organized in a separate union. Until 1966 the organizing efforts had concentrated on net fishers (gillnetters and seiners). In expanding beyond these two gear types the UFAWU became embroiled in disputes with a number of other organizations. It fought with the Deep Sea Fishermen's Union (DSFU, a CLC affiliate) over representation of longline crews sailing out of Prince Rupert. Many of the vessel owners in this fishery were also members of the Prince Rupert Fishing Vessel Owners Association (PRFVOA). The Prince Rupert Fishermen's Co-op also had a large number of big vessel owners who belonged to this association and entered the dispute with the UFAWU, a dispute which almost tore apart the community of Prince Rupert. Shoreworkers found themselves in the middle of the conflict that ensued.

SHOREWORKERS SPEARHEAD THE PRINCE RUPERT STRIKE

The ensuing strike hinged on the refusal of shoreworkers to handle fish caught in unorganized bottoms. This was precisely the reason shoreworkers had been organized in an industrial union in the first place, and the outcome of the battle lay with them. On 23 March 1967 the Supreme Court granted the PRFVOA an injunction against Prince Rupert shoreworkers who refused to handle "hot" fish. A coastwide UFAWU ballot was held on 3 April, and a decision was made not to send a telegram ordering Prince Rupert shoreworkers to handle "unfair" vessels. The strikes were justified on the basis that the companies were expanding into other major fisheries. "Refusal by the Association [PRFVOA] to negotiate a trawl agreement and the abrupt lockout of longliners are being linked with the major expansion of the B.C. trawl fishery, reflected in a better than 100 percent increase in landings in the past three years, and plans by major monopolies in the industry to enter the trawler field on a large scale" (*TF* 14 April 1967, 12).

Vessel owners exploiting the groundfish industry organized the PRFVOA and they also expanded operations into the net fisheries by constructing a salmon cannery the Co-op owned. Just as the big fishing companies were expanding from salmon canning to other fisheries and processing techniques, so too organized large vessel owners in the north were expanding from the groundfish industry into the older established salmon fishery and the salmon canning industry. The UFAWU felt simultaneously challenged on two fronts.

As mentioned, the brunt of the strike fell on Prince Rupert shoreworkers. The UFAWU had no effective way to prevent big vessels from fishing. The only way to block them was to refuse to unload their catches at the plants and to refuse to supply them with ice and bait. This proved to be extremely effective. However, it also polarized the community. Wives of big vessel owners formed their own organization called the "Marching Mothers." They held parades and generally waged a campaign against the union. The female membership of the UFAWU in Prince Rupert issued the following statement in response: "We are the Allied part of the letters [UFAWU]. Some of us are wives of fishermen and all of us are shoreworkers ... Are the 'Marching Mothers' shoreworkers?" These women went on to note that they were receiving a fair wage, "thanks to our Union," although they were not being overpaid. Women had to stand ten to twelve hours a day, filleting, working on large halibut. "But when our fishermen need our help as they have helped us in the past, we intend to give it to them ... At the same time, we are not going to turn our backs on our fishermen who bring in the fish we process; or on the Union which has helped us to have the working conditions and wages we enjoy today" (*TF* 30 June 1967, 5). Here was a clearly articulated instance showing the intersection of class and gender relations. On one side were the "Marching Mothers," who were allied with the small-time capitalists – the big boat owners, who also owned and operated the Co-op cannery. On the other side were the female wage workers allied with the union fishers. What made this battle so emotionally charged was the taking of issues beyond the workplace, locating class cleavages within the community of Prince Rupert itself.

Race also became an issue when employers threatened with deportation those shoreworkers who were recent immigrants or foreign born. On 24 May 1967, a Co-op worker issued the following broadcast over a local radio station: "UFAWU trawlers and longliners have declared Rupert Vessel Owners' boats unfair, therefore, we won't touch them or any other 'hot' fish coming from them" (*TF* 26 May 1967, 4). On 2 June, shoreworkers closed down the entire Co-op plant, declaring it to be unfair. When Co-op and PRFVOA vessels then tried to unload their catches at B.C. Packers' Seal Cove plant and at the Royal Fisheries' plant, workers staged plant walkouts. The UFAWU set up pickets at the Co-op plant. The DSFU retaliated by picketing the Seal Cove and Atlin plants, declaring the UFAWU to be unfair. Injunctions were invoked. Jack Nichol, the union's business agent who had replaced Alex Gordon as shoreworker organizer, was arrested twice. The UFAWU president and secretary treasurer were eventually jailed for ten months. Vessel owners launched legal suits against the union amounting to a quarter of a million dollars in damages.

The dispute ended in July when the DSFU and the UFAWU worked out a jurisdictional agreement. However, the strike ended with no negotiations on behalf of the shoreworkers who had supported the strike and who had refused to cross picket lines. Eight employees at Royal Fisheries were rehired after a thirty-day "cooling-off" period, and their seniority was restored in early September. However, irreconcilable differences had developed at the Co-op plant. Approximately fifty shoreworkers there supported the strike and the UFAWU attempted to restore their seniority rights. Co-op management demanded a thirteen-month "cooling-off" period during which time seniority would not be recognized. This meant that they would lose employment during the fall, winter, and spring. Eventually a compromise was negotiated. Ten of the displaced workers were to be reinstated in January 1968, ten more in March, and the remainder would be called from the list of the "outside group" in the order of their original seniority. They were to be fully instated as of 1 September 1968, after losing their seniority for the summer months. The formula was never implemented.

In February 1968, 104 Co-op shoreworkers and office employees voted to decertify the UFAWU. None of the fifty suspended workers were allowed to vote. The Co-op kept the most militant workers out of the plant and replaced them with employees who were intimidated by the threat of similar action. In a similar vote Royal Fisheries decided to keep the union. The DSFU took over from the UFAWU at the Co-op and created what amounted to a company union, the Prince Rupert Amalgamated Shoreworkers and Clerks Union (PRASCU). On 6 November 1968 Co-op employees waged a six-hour strike, demanding that the seniority rights of the "outside group" be cancelled. They won their demand. The divisions among Co-op shoreworkers were based on support of vessel owners (some shoreworkers were related to them) versus support for the union. Both groups supported fishers but were divided along class lines, in terms of class interests stemming not from their own conditions of work but from those of the fishers with whom they were allied. In the end, shoreworkers were the real losers, since neither the Co-op nor the UFAWU took much action to help them once the fishers' strike had terminated. However, the strike did provide an important learning experience in teaching shoreworkers militancy, even if it had been in someone else's war.

The experience of PRASCU serves to highlight the importance of an industry-wide union in the fisheries. Most of its agreements have mirrored those between the UFAWU and the Fisheries Association. However on 23 June 1978 the Co-op locked out its 500 employees. The main issue was the protection of seniority rights. The Co-op offered an extra 10 cents per hour increase over the general industry settlement providing that employees agreed to the replacement of departmental seniority with seniority based on job function. The Co-op sought injunctions against the use of pickets at its Prince Rupert and Vancouver plants. Some of the employees at the Vancouver plants were UFAWU members and thirty of them respected the pickets, but much hard feeling developed between members of the two unions. The B.C. Labour Relations Board upheld the right of PRASCU to picket on the grounds that the union had little or no leverage apart from this strategy. The strike ended seven weeks later when shoreworkers narrowly voted (58 percent) to accept the proposal they had already rejected three times. The Prince Rupert press attacked PRASCU's business agent, causing her to resign after five years of service. A union formed within one company was vulnerable to considerable pressure from the employer and was virtually ineffective in an industry like fishing. The history of union organization, or rather the lack of it, among shoreworkers in the Atlantic fisheries, where laws were enacted preventing extensive regional representation, supports this conclusion.

"The United Fishermen and Allied Workers' Union has set the tone for labour activity on the West Coast since its establishment in 1945 ... This union has made itself the mark against which all other organizations in the industry must be measured" (Clement 1986, 83). Clement goes on to quote from John Boyd's report in 1968 to the Task Force on Labour Relations (also known as the Woods Commission): "the Pacific Coast fishery is the only one with significant organization ... The Atlantic Coast Fishery and the Inland Fishery are almost entirely unorganized, especially at the level of actual fishing operations" (83).

In trying to widen its jurisdiction, the UFAWU almost destroyed itself. While conflicts among fishers continued to create divisions among them, shoreworkers became an ever more cohesive body within the union. They finally came together to struggle over the issue of sexual discrimination in wages and job classifications.

THE GENERAL STRIKE OF 1973: SHOREWORKERS DEMAND AN END TO SEXUAL DISCRIMINATION

As early as 1954 the provincial government had passed legislation promoting the principle of equal pay for work of equal value, but the Fisheries Association flatly refused to incorporate the principle into the agreement that year. Like similar pieces of federal legislation enacted during the war years, the provincial bill was ineffective because it carried no provision for enforcement. In 1968 Mickey Beagle, one of the general organizers working for the union, presented a brief to the hearings of the Royal Commission on the Status of Women, in which she outlined the cumulative effects of the years during which companies used women as a pool of cheap wage labour. An experienced female general fish worker's wage was 9.3 percent less than that of an inexperienced male worker, and 24.5 percent less than an experienced male employed in the same category. A fully qualified filleter received $2.34 an hour and she had to pass tests as well as meet production standards to earn this rate of pay. Meanwhile, inexperienced men received $2.37 an hour. Helen O'Shaughnessy, vice president for shoreworkers, noted: "Women generally perform much of the key production line work ... and are affected directly by speed-up and intensification of work loads." The Royal Commission called for stronger provincial equal pay legislation with provisions for enforcement. A clear definition of what was meant by "equal pay" was also necessary, since work women performed could be entirely different from work done by men but still be equal in value; for example, it could be equal in terms of training time, skills, hours of work, productivity, intensity, and necessity to overall production. "The difference in work performed by men and women is not in the skills [but in] the fact that there is some heavy work performed by men that women are not physically able to do. This does not mean that work performed by women is not equal – it simply means that at times it is not the same" (*TF* 3 May 1968, 9).

In the late 1960s women became more aggressive in demanding equality in employment and pay. Many of them were losing their jobs. The 1969 closure of Klemtu, on the Nass, in addition to further closures of plants on the Fraser and Skeena rivers, resulted in the loss of an estimated 1000 jobs. The twelve canneries operating in 1968 were reduced to five by 1971, with a loss of 1600 jobs. In 1968 the salmon cannery at Namu was closed. This had been the only cannery left in operation on the coast between the lower mainland and the northern area. The village of Bella Bella was deprived of its major source of employment, a fate it shared with the village of Klemtu. Most of these plant closures affected aboriginal shoreworkers and fishers more severely than other groups.

As aboriginal women became more militant, they also became a less attractive labour force for employers. Women had replaced Chinese labourers when the latter became more militant and scarcer in terms of numbers. With diversification, new jobs were created and staffed by women, while many of the original tasks were rendered obsolete with mechanization. In the 1970s this process was repeated with the creation of a new fishery: roe herring.

In the 1960s Japan, having overfished its own stocks, was looking for new sources of roe supply. Fishing companies in British Columbia encouraged Japanese interest in provincial waters. Roe was already being extracted from salmon, but a really lucrative market lay with herring. Since gillnetters and seiners already captured salmon and herring, existing gear types could be used. The innovation as far as labour was concerned involved shore plants, where additional lines were established to pull the eggs. Since herring runs preceded salmon, seasonal employment for women could be extended, and this became the source of the ensuing conflict.

In the early 1960s Japanese capitalists developing this fishery in the province began to import labourers from Japan. The process began when forty Japanese technicians began to work at the Coal Harbour whaling station, jointly owned by B.C. Packers and a Japanese firm. The companies claimed that Canadian workers were not skilled in butchering whale meat for human consumption. The UFAWU charged that these skills could easily be taught to Canadian crews. While this particular operation was short-lived, it set the pattern for other processing operations with Japanese investment. For example, labourers from Japan were employed in the sea urchin roe fishery on Vancouver Island. But the conflict came to a head over the lucrative roe salmon and roe herring fisheries. In 1969 the UFAWU applied to the federal minister of manpower and immigration to prevent the import of workers from Japan into salmon roe operations. The application was ignored as even larger numbers of workers were brought to work in the herring roe fishery. Between 1971 and 1978, an estimated 400 working permits were granted. Union members suspected that these labourers were paid much lower wages. The UFAWU saw the use of foreign crews as a means of trying to introduce a cheap and unorganized labour force into a new industry. And as indicated in chapter 4, this development could also be placed in the older context of the use of cheap wage labour under the contract system. This time, the capitalists were themselves from Japan, reflecting the multinationalism now prevalent in the fishing industry (B.C. Packers under Weston was also processing in other countries), but the use of cheap wage labourers continued apace.

Another tactic the companies used to keep wages as low as possible was to substitute women for men. "Until a few years ago salmon eggs were regarded as 'waste' and discarded with the rest of the offal. Whenever the eggs were needed for bait they were pulled out of the fish by men" (*TF* 26 Jan. 1971, 6). Once a large market for the roe was created, however, only women were assigned to these operations. Employers claimed that women's fingers were more agile than those of men, and that women were therefore more suited to the work. The women asked for the same rate of pay as the men had received for pulling eggs. The companies refused. The women, although judged to be more suited to the work, received wages four cents above the base rate paid to women, or 34 cents an hour less than the rate men received for this work (6).

The roe herring operation expanded in 1972 when eighty-five boats participated in the new fishery. A number of fishers made fortunes overnight, and the companies realized substantial profits. But shoreworkers were left out of this new "bonanza." While a number of women received several weeks of extra work, the processing of herring roe on assembly lines was introduced during the term of an existing agreement. The companies refused to negotiate wages or conditions for the new operation in either the 1972 or 1973 seasons. In the meantime herring fishers won a price increase of 140 percent, while herring tendermen won substantial wage hikes. This injustice was a major factor fuelling dissatisfaction among shoreworkers. The nature of the operation compounded the conflict. Since it was a largely manual process, it could easily be contracted out to small unorganized firms, and these seemed to mushroom overnight. Some of the larger established companies farmed out this operation to them. The UFAWU called on fishers to deliver their catches only to unionized plants.

By 1972 women workers were also feeling the impact of negotiations carried on over a number of years based on percentage increases. That year women earned an increase of 3 cents an hour lower than that received by men, and women working in fresh fish plants received an increase that was 6 cents lower. The base rate for women in 1972 was $2.89 while men's base rate was $3.87 per hour (*TF* 21 Jan. 1972, 4).

By 1973 all the aforementioned factors combined to make shoreworkers militant. Jobs remained a primary consideration. Japanese demand created a lucrative fresh/frozen market, diverting salmon from the can to the freezer, thereby eliminating many jobs traditionally assigned to women since freezing was less labour intensive than canning. New operations, like popping roe, which were designated women's work, carried correspondingly low rates of pay. Where feasible, operations were contracted to unorganized firms that used even cheaper labour. For example, in 1975 a plant in Tofino, on Vancouver Island, tried to export frozen herring to Mexico for processing. Sixty UFAWU shoreworkers were laid off when they refused to load 250 tons of the fish into freezer trucks. A Japanese firm which had investments in this plant had also purchased one in Mexico, where hourly wages were $1.20, compared to $4.50 in Tofino. The union succeeded in having legislation enforced that prevented a company from exporting more than 25 percent of its total landings in a raw form (*TF* 16 May 1975, 12). Meanwhile, CANFISCO was financing research into techniques that would allow herring roe to be processed mechanically.

Women employed in the fresh fish sector were also dissatisfied. The expansion of B.C. Packers into Atlantic groundfish led to a policy whereby it only operated in this area on the Pacific coast if substantial profits could be made. Women employed in fresh fish operations, who had until then had the most secure employment, were now subject to sporadic layoff. This was in direct contrast to the situation of fishers which was greatly enhanced by the opening of the Japanese market to provincial fish products and the establishment of the 200-mile limit. In 1973 shoreworkers had had enough and went out on strike. "The union proposes to delete all contract references to women and simply list classifications and the pertinent wage rates. As a union whose shoreworker membership numbers women in the majority, the UFAWU has a responsibility to establish a lead in a fight for equal rights for working women and equal pay for work of equal value" (1973, 30 March, 12).

The central issue in the 1973 negotiations was not wages but rather the introduction of one cannery schedule. The Fisheries Association agreed to merge the two cannery supplements, to list job classifications by labour groups, to standardize rates, and to raise the lower paid classifications by 15 cents over a period of two years, in addition to any general increase in hourly rates. But it proposed to do this by paying men lower wages than those they currently earned. "Clearly, the union's aim is to raise the level of the rates paid to women and not to depress male wage rates ... Discrimination against women is a fact of shoreworker contract life, prompting a Victoria newspaper recently to cite fishing industry agreements as clear examples of practices that might contravene human rights legislation" (*TF* 27 April 1973, 11).

The number of shoreworkers attending union meetings was larger than it had been in decades, and the negotiating committee received the strongest support from its membership in the history of the union. Cannery, net, reduction, and watchmen classifications voted 90.5 percent in favour of strike action, while the fresh fish and cold storage sector voted 92.6 percent for a strike. The reversal in strategy in favour of shoreworkers compared to fishers and tendermen was duly noted. "In the 28-year history of the UFAWU, shoreworkers have been on strike a total of five weeks. Even when they have struck – as in 1959 and 1963 – their actions have been closely linked to tieups of salmon net fishermen" (*TF* 22 June 1973, 8). This time they were on strike to support demands to better their own conditions of employment. The resulting general strike lasted only one week (from July 6 to 15), with shoreworkers and tendermen becoming more militant as time went on compared to fishers who capitulated.

While a breakthrough was realized as a result of this strike, inequalities among shoreworkers were not totally eradicated. An hourly differential of 71 cents between filleters and male labour rates was narrowed to 21 cents. References to male and female networkers was abolished, and as of 15 April 1974 rates were standardized. In cannery classifications, the hourly base rate differential of 44 cents between men and women was eliminated over a period of three years. The new wage structure elevated B.C. shoreworkers to among the highest paid workers in a primary food industry in the country, but differences based on gender were only narrowed, not eliminated.

CONCLUSION

Salmon canners originally created the structure of labour and wage conditions among shoreworkers by differentiating between groups on the basis of race and gender as well as by using existing ties to pre-capitalist relations of production to create cheap wage labour. Until the Second World War, employers dictated conditions to plant workers. Although the historical militancy among fishers resulted in their forming a province-wide industrial trade union, made possible by changes in federal legislation that rendered the contract system obsolete as a method of paying labour, being in a trade union did not mean that shoreworkers participated on an equal footing with fishers, machinemen, or tendermen. Until the strike in 1973, fishers dominated the union, while shoreworkers slowly came to develop leadership skills within the union structure and to articulate issues of concern to them. But to be successful shoreworkers also had to develop the militancy that had characterized the conflicts of fishers since the turn of the century. They had to overcome exploitation stemming from their conditions as cheap wage labour as well as the patriarchal legacy of oppression of women and the colonial heritage of racism against "non-white" groups. Aboriginal women experienced all these dimensions in a particularly acute form, especially in light of the fact that their wage labour also pitted them against those of their male relatives who had considerable capital investment in the fisheries and in turn sought to employ them as cheap wage labour in, for example, fish plants and canneries operated by First Nations associations. In the 1970s the shore sector emerged as a cohesive and militant group within the union, while fishers became increasingly divided.

In 1977 Jack Nichol replaced Homer Stevens (who did not run for re-election that year) as president of the UFAWU, winning by acclamation. Nichol had started as a shoreworker and had been a shoreworker organizer since the mid-1960s. In the 1970s provincial shoreworkers were said to be the highest paid fish workers in the world, receiving higher wages than their counterparts in the United States. For example, in 1983 women in the classification Egg Pullers, Packers, Sorters, and Slimers received $6.46 per hour in plants located in Bellingham and Puget Sound in Washington state (according to an ILWU contract). The lowest rate, apart from probationary employees, received by UFAWU members in 1982 was $11.18 per hour, almost double the rates received across the border.

While hourly rates were impressive compared to those in other similar industries, especially to the wages Atlantic fish workers earned, seasonal employment translates into low annual incomes. Guppy has provided empirical data for the period 1975 to 1983 on wages, numbers employed and conditions of employment (for example, number of weeks of employment per season) for shoreworkers. For example, annual fish plant earnings for Indian women were $9000 compared to the $20,000 earned by Anglo-Canadian European men (Guppy 1987, 220). Conditions of inequality based on racial and gender distinctions clearly persist. But improvements have been made, and they have been made by shoreworkers working together and using the union to press for their demands.

Unfortunately, companies have many ways to respond to militant workers, as is evident in the severely decimated ranks of plant workers. Ultimately, companies can simply relocate, a move made all the easier when the companies themselves have been bought by huge food conglomerates like Weston, which owns not only B.C. Packers but also Connors Bros., Weston's maritime fishing industry subsidiary (Meggs 1991, 213). The fishing strike in 1980 "triggered a reorganization in the processing sector which left Weston's B.C. Packers completely dominant among the survivors" (211). Meggs (219) also notes that in the space of three years following the crisis in 1980, a thousand shoreworkers lost their jobs. At the end of chapter 4, I provided some of the impact of such cuts on aboriginal female shoreworkers since the Fish and Ships research project took place in the years immediately following this crisis (it was in June 1982 that we travelled to Prince Rupert).

In the late 1980s the Americans also successfully invoked the GATT agreement to challenge the practice of retaining raw unprocessed fish in British Columbia rather than exporting it for cheaper processing elsewhere. The success of the UFAWU in promoting the interests of shoreworkers has hinged on this condition, for which it has fiercely fought over the decades. Finally, the few remaining manually intensive operations that require large numbers of labourers are being mechanized (for example, fish boxes are being introduced to clean fish, a task traditionally assigned to women, especially aboriginal women). But what the history of the UFAWU serves to demonstrate is that the groups that come to be designated as cheap wage labour are by no means a passive labour force. They have been creative in the ways in which they have challenged exploitation of their labour power by employers. Trade unions like the UFAWU have been only one means of resistance.

Gillian Stainsby based her 1991 thesis in part on her own experience as a shoreworker. She interviewed twenty-three women shoreworkers from Prince Rupert, Vancouver, and Steveston. While women continued to be highly paid ($13.55 in 1988 for most women), Stainsby argues that the seasonality prevalent in the industry erodes the potentially high incomes and forces women to rely on government-sponsored social net programs like unemployment insurance. Her description of job categories suggests that many of the trends identified in this chapter have persisted. She notes, for example, that "the most highly paid union workers are the boatshop tradesmen" paid monthly, and all the monthly paid workers of whom she was aware were men (Stainsby 1991, 49). She calculates that in 1985 women in B.C. earned 77 percent of what men earned in the fish processing sector. "Unless they work full time, women shoreworkers' incomes are below the poverty line" (53). And she focuses on women's double day of labour, incorporating into her analysis the work women perform in the home as well as on the factory floor.

Technological changes introduced in the latter part of the 1980s included the herring sexer (used in processing roe) and the coastline fish-washing machines used in canneries and fresh fish departments (Stainsby 1991, 100). The resulting effects on the women who operate them are the same as those noted earlier in this chapter. Women lose control over the labour process and are forced to work to the speed of a machine. Finally, Meggs (1991) also notes the dislocations caused by the free trade agreements among Canada, the United States and Mexico which have jeopardized the GATT agreements as regards the export of fresh fish unprocessed to the U.S. Another blow to organized fishers is an ever increasing reliance on ocean ranching to produce farmed fish for processing and consumption. These developments are but the latest in an industry dominated by the profit motive and the ongoing efforts of processors to produce the commodity (now extended to the raw resource itself, the fish) as cheaply as possible. In turn, there are obvious parallels here to the creation of cheap wage labour forces in other Canadian industries and around the globe. As British Columbia shoreworkers began to be paid decent wages, employers simply shifted their operations elsewhere, mechanized existing plants even more intensively, and put pressure on the federal government to allow the flow of the unprocessed commodity across the border where it could be handled more cheaply in subsidiaries of large Canadian conglomerates like Weston.

6 State, Labour, and Capital

There is a vast literature on the role and place of the liberal democratic state in the capitalist mode of production. Most contemporary political theorists include Marxist analyses as one of several theoretical approaches. Since the famous Miliband/Poulantzas debate, there has been a burgeoning of Marxist writing on this topic, although there is no consensus on how to conceptualize the state within advanced capitalism. Marx himself had planned to write a volume on the state, but only managed to complete the first book in the series, *Das Capital* (Foreword, *Grundrisse* 1973, 54).

Held contends that at least two strands in Marx's work relate class to the state. The first emerges most clearly in Marx's early work and treats the state as "relatively autonomous" from the ruling class or classes. The second position, which Held sees as prevalent in Marx's later work, treats the state as more or less an instrument of the dominant class interests, and involves a narrowing of the first position (1989, 33).

Marxist theorists have picked up and elaborated the second position in the base/superstructure debate. As indicated in chapter 2, this has resulted in the charge of various "isms": reductionism, determinism, functionalism, economism, etc. To view the state as a mere superstructure that emerges from the economic conditions of capitalist class exploitation also replicates the dichotomous thinking criticized in that same chapter. However incomplete Marx's own writing on the subject, the concept of relative autonomy facilitates an understanding of the creation of the B.C. state and its involvement in the fisheries, which was by no means tied directly to the interests of salmon canners. Various pieces of legislation were passed year after year in the late nineteenth and early twentieth centuries – only to be nullified by Ottawa, generally to conform to international relations as determined by Great Britain – which tried to restrict and/or exclude entry into British Columbia of Chinese, Japanese and East Indian immigrants. Examination of this legislation makes clear that racist xenophobia was not always in the best interests of employers and that politicians represented and gave voice to a very mixed assortment of constituent viewpoints and interests (Roy 1989; Ward 1978). The role of the federal state, since it controlled ocean fishing and was responsible for "Indian affairs," further complicates the web of relations among politicians, government bureaucrats, salmon canners, and their labour forces.

In addition, the formation of organizations like the NBBC, the organized union movement within the fisheries (the UFAWU), and the associations formed by salmon canners (the SCOC) and later broadened to include all major processing companies (the Fisheries Association) further complicates the manner in which the state has participated in the fisheries.

Also at issue are the jurisdictional boundaries of the two levels of government, provincial and federal. For example, in his attempt to establish a floating cannery in 1924, Francis Millerd challenged the authority of the federal government to issue cannery licences. "On 23 September 1927, a judge of the B.C. Supreme Court confirmed Millerd's acquittal, wiping out federal authority to control cannery licensing ... Without the power to licence canneries, Ottawa could not limit entry to the processing sector or limit Asian ownership of canneries, salteries or fishmeal plants" (Meggs 1991, 138). At this point the salmon canners took matters into their own hands. Various delegations went to Ottawa and the result was the creation, on 18 May 1928, of a new B.C. Packers controlling forty-four canneries. Millerd Packing was added three days later to the corporation's assets. According to Meggs, the federal fisheries minister and senior officials in the department stood by and allowed the corporation to institute its own plans. "Under the terms of a contract agreement, which included canners handling more than 80 per cent of the province's salmon production, the coast was divided into seventeen areas. Within each zone, the canners established strict limits on the numbers of seines, gillnets and drag seines allowed to operate" (139). Thus, in trying to establish greater autonomy from federal government control, in the end Millerd's actions proved suicidal since they allowed the largest company to step into the breach and acquire the dominance it had seen eroded since its previous takeover in 1902.

Both the UFAWU and the fishing companies have lobbied Ottawa or had recourse to existing legislation in their ongoing battles with each other. In fact, the creation of the welfare state after the Second World War led to the emergence of the UFAWU, since changes in labour legislation permitted the formation and recognition of an industry-wide trade union. Until the legislation enacted by a welfare state enforcing minimum wage legislation and social support services like unemployment insurance and pension schemes, very little legislation protected shoreworkers. "Between 1940 and 1948 thirteen new federal welfare statutes were passed ... Nine of the statutes dealt with veterans' benefits while the remaining four – the Unemployment Insurance Act 1940, the Family Allowances Act 1944, the National Health and Welfare Act 1944 and the National Housing Act 1944 – laid the ground work for the welfare state" (Ursel 1992, 216–17). Garrod (1984, 15) notes, for example, that while the "Female Minimum Wage Act" of 1918 was established to cover women working in a variety of industries, a decision made in 1919 excluded females working in the canned fish sector of the fishing industry. And regulation of wages in fish processing other than canning applied only to "women specially imported from the Old Country" and local women trained by them. "Thus it appears that the hardships and skills associated with this work were only to be compensated for white women, Indian and Japanese women remained unrecorded and unprotected" (15–16). The contract system enabled the bulk of these labour forces to be hired and rendered invisible since no accounting to employers and thus to government officials was made necessary. This is also why it becomes so difficult to provide details on their numbers and wages since no official accounting was required until after the end of the Second World War. "It is clear that salmon cannery workers, mostly Asian and Indian women and men, were a super-exploited class of workers, with wages and conditions substantially lower than 'white' races, and with no state regulation of this exploitation. Even into the 1940s, the salmon canning industry was organized in much the same fashion as at its beginnings in the 1870s and 1880s" (20).

THE QUESTION OF ABORIGINAL GOVERNMENT

The creation of the B.C. provincial state in 1871 marked the formal and final transition from colonial rule and the fur trade to European settlement and indigenous industrialization. The formal negotiations to incorporate British Columbia as a Canadian province did not take into account the political structure of aboriginal peoples and the form of government they had developed to hold together the various tribes and nations along the northwest coast (Fisher 1980). However, the federal Canadian state was quick to outlaw institutions like the potlatch which were central to aboriginal government when it assumed federal control of "Indian affairs." In legally banning the institutions that politically unified the various tribes, the federal state created a void which enabled both it and the provincial state to alienate land and resources previously under the collective control of the various tribes without negotiating the transfer in a way that recognized aboriginal political systems. Ignoring Indian claims to land and resources was facilitated by the enormous decimation of the population that had taken place over the last century. While the aboriginal population in the area that became British Columbia was estimated at 86,000 in 1780, by 1885 it had decreased to 28,000 people (Duff 1977, 39).

When it came to disputes over fishery sites, the colonial and provincial governments were not adverse to sending armed patrol boats to aid salmon canners in taking sites and establishing canneries on them, especially in the north where white settlement remained scanty until the establishment of the city of Prince Rupert and where the First Nations remained militant, especially concerning their right to traditional fishing sites. The evidence gathered by the Metlakatlah Commission of 1884 is especially instructive in providing details of armed struggles between patrol boats dispatched by the provincial government on behalf of salmon canners and Indian villagers seeking to retain control over fishery resources. Missionaries entered these disputes and Duncan established his own settlement of Metlakatla, recruiting aboriginal peoples from the area. He became involved in trying to protect his settlement against the encroachment of salmon canners and white settlers, thereby running afoul of the provincial government.

The establishment of the liberal democratic state involves the consolidation of private property rights necessary for capital accumulation. In 1763 King George III issued a proclamation in effect recognizing Indian rights of possession to the lands they had occupied and used. This proclamation had the force of a statute in the colonies, and the federal Canadian state was charged with its enforcement, meaning that lands had to be ceded voluntarily. The means of alienating these lands was the treaty. However, the area that became British Columbia, apart from a small portion of the north-eastern section of the province, has never been ceded by treaty. Douglas also purchased land on Vancouver Island around Fort Victoria and at Fort Rupert in 1850 and 1851, but Tennant (1990, 19) emphasizes that the word "treaty" is not used in the documents nor was it used by Douglas, although the Canadian courts later ruled that they constitute treaties.

In general, however, the colonial government of Vancouver Island and of British Columbia "evolved a policy that ignored or denied the existence of any native title and therefore any need to make treaties" (Duff 1977, 67).

Thus the policies of the provincial government involved a denial that the aboriginal population had any rights to pursue its traditional economic activities. There was also a popular misconception that Indians had no system of government of their own, and politicians made use of this ideology to establish and extend private property rights over land and resources. This misconception also appears in political theories of the state. A theoretical reformulation thus requires a reappraisal of those theories that begin with an original assumption of a "state of nature" counterposed to the creation of the state (and civilization). As we saw in chapter 2, Marx himself was not free from such misconceptions.

POLITICAL THEORIES AND THE "STATE OF NATURE"

In summarizing the history of Western political theories of the state, modern analysts tend to begin with Hobbes, Locke, and the two Mills. For example, Held (1989, 12) identifies four strands of such thought: liberalism, liberal democracy, Marxism, and political sociology. His starting point is liberalism and the writings of Hobbes. Knuttila (1987), in his survey of theories of the state, also begins with classical liberalism; that is, the political theories of Hobbes, John Locke, the utilitarians Bentham and James Mill, and John Stuart Mill.

In *Leviathan*, Hobbes (1588–1679) develops four interrelated concepts beginning with the "state of nature" and incorporating "right of nature," "law of nature," and the "social contract." He envisages the "state of nature" as the situation in which individuals are without a "Common Power" or state to regulate social behaviour. The result is the "Warre of every one against every one." In this state the life of the individual is "solitary, poor, nasty, brutish and short." To end this situation, individuals need to recognize an interest higher than themselves and submit to its authority – the concept of the modern state. Held (1989, 16) summarizes Hobbes' argument as follows: "If individuals surrender their rights by transferring them to a powerful authority which can force them to keep their promises and covenants, then an effective and legitimate private and public sphere, society and state, can be formed. Thus the social contract consists in individuals handing over their rights of self-government to a single authority – thereafter authorized to act on their behalf – on the condition that every individual does the same. A unique relation of authority results: the relation of sovereign to subject."

According to Knuttila (1987, 20), Hobbes would argue that "the existence of a sovereign state represents an essential precondition for the existence of a civilized human society; any human intervention beyond that of continual war against all requires a state. The state is thus at the core of human existence." While Locke (1632–1704) did not begin with the same premiss, he also posited a "state of nature" built this time on the exercise of reason and private property. Locke made an argument for a franchise based on property holding, although "women and slaves of both sexes" were disqualified altogether (Held 1989, 21). Later political theorists took up and modified these arguments, while Marx located a major flaw in their focus on the negotiation between the individual and the state, unmediated by the reality of class struggle.

The problem with using Hobbes and classical liberal theory as a starting point is that the idea of the creation of the state is located in a void. The concept of a "state of nature," however it is defined, persists even in the writings of Marx. Thus a theoretical understanding of the modern state is assumed to be counterposed to a state where the forces of nature rule men, causing them to organize society in direct response to their physical need to survive. In other words, the concept of the state is dichotomously posed against its negation ("state of nature") with the ideological connotation of the first as "good" and the second as undesirable.

Missing in these theories is the recognition that the liberal democratic state in western Europe was itself the product of a long series of struggles, which could be traced to the period that led to the founding of the city-states of the classical Greeks. Also missing is the fact that humans govern themselves in complex ways without necessarily having to create a state formation. Indeed from this perspective, the creation of the state does not liberate the individual ["man"] but rather confirms his unequal position in society and at the same time denies any rights whatsoever to whole groups of people, including women, children, and slaves. The extension of the rights of citizenship to these groups then occurs in struggles of various sorts within the framework of the liberal democratic state. Thus the ideological notion of the positive features of liberal democracy are, in an odd way, preserved.

What is missing from these analyses is the premiss that the state was founded not in a state of nature but in a state of violence, and that the emergence of the modern state consolidated class domination, patriarchal authority in the subjugation of women and children to men, a state monopoly over armed force, and the rights of private property as an economic base for public participation in politics (Parkin 1982, 71–89). Although Max Weber, among others, recognized that the formation of the state is based on coercion, he also missed the implications of the struggles that preceded the establishment of the state, such that the factual reality and necessity of the state in modern society goes unquestioned. A recent example of this underlying assumption of the "naturalness" of the state is the standoff in the summer of 1990 between the Mohawk Confederacy at Kahnesatake and the Canadian armed forces sent in to replace the Sûreté du Québec. A central issue was the right of the Mohawks to govern themselves independently from the federal and provincial states, including their right to arm themselves. That a system of government is even possible without the framework of the nation state is a foreign idea to most Canadians but one that has been placed on the agenda in the issue of Quebec's sovereignty and the rights of aboriginal peoples living within the current boundaries of Quebec to self-determination.

By situating the emergence of the Western state in its historical context, the dialectical interplay between production for use and production for exchange emerges in its various possibilities. Of particular interest in this context is the collection of essays edited by Coontz and Henderson (1986). These essays deserve a brief review because they indicate the type of reappraisal that is necessary in theoretically situating the emergence of the liberal democratic state in western Europe – and its transplantation to North America – from much older political and state formations.

THE EMERGENCE OF THE CIVIL STATE AND THE DESTRUCTION OF KINSHIP-BASED SOCIETIES

The contributors to *Women's Work, Men's Property: The Origins of Gender and Class* (Coontz and Henderson, 1986) share certain key assumptions. One primary assumption is the need to historically situate the establishment not only of male dominance but also of a gender-based division of labour. In searching for the roots of these two phenomena as social constructions the authors reject the notion that either is biologically given. That is, they reject any formulation that situates women's subordination as somehow rooted in human nature.

Leibowitz (1986) examines hominid societies to trace the beginnings of production. "Production appeared when individuals or groups deliberately set out to collect goods to give (or distribute) to others ... Carrying food back to share with others is production" (70). Thus her analysis merges from the outset production for use with the exchange of the products of individual labour with others. She calls this exchange with a small "e" to distinguish it from a later form of Exchange with a capital "E" that involves Exchange among groups rather than exchange among members of one's own group. Important in this context is the further point that kinship as a way of socially organizing human society is a social construction that may have developed relatively late. "Kinship is not the only, nor even necessarily the primary, means of establishing interconnections between people" (Coontz and Henderson 1986, 116). While exchange draws the members of a group together in communal relations, Leibowitz suggests that when groups begin to specialize in order to produce goods for Exchange with other groups, the possibility of inegalitarian relations is present, although by no means inevitable.

Leibowitz also argues that although hominids were more physically dimorphous than their descendants, we cannot assume male dominance among them. High mortality rates meant that many hominids did not live far into their adult years, creating groups with a large proportion of children among whom sexual and other physically differentiating characteristics would not be highly developed. The burden of providing food would include these children, who could become major food providers depending on the demographics of the group. Thus a division of labour by either sex or age must be explained rather than being assumed as given in human nature. Other contributing authors dispute Leibowitz's contention that a gender division of labour was already established with hominids, dating it much later in human history. In either case, however, division of labour by categories such as age or sex needs to be proved rather than assumed.

The other contributing authors examine the growth and development of various lineage societies, and some then trace the further development from these of kingdoms and archaic states. Thus the civil state is seen to emerge from a long line of development that features highly complex social formations. The authors also stress the variety of lineage societies. While they all posit initial formations that were matrilineal and matrilocal, once established, further variations became possible, including the emergence of patrilocal lineage organizations.

Lineage societies (which the authors establish as a distinct mode of production, including in their examples some of the aboriginal tribes living on the Pacific Northwest coast) have a developed gender division of labour whereby the women control production for use within the lineage while men come to specialize in the Exchange of products with other lineages. As long as these remain matrilocal, relations between men and women tend to be reciprocal. It is only with the emergence of patrilocality that the *possibility* of subordinating women to men emerges, since the men in these lineages need to have access to the products of women's labour, especially in arranging the feasts necessary to Exchange with other groups.

While access to this type of production is assured men in matrilocal societies, it becomes problematic in patrilocal lineages. With patrilocality, men come to realize there is an advantage in individually controlling the labour of a number of women, since this then gives individual men an advantage in Exchange. Once established, patrilocal lineages come to see a further advantage in dominating other groups, and thus patrilocality is forced upon other social groupings through external coercion. "Women were already producing surpluses, but were doing so freely. Certain men could well appreciate what they might win by forcing – for the first time in human history – certain human beings to work for others ... It is enough for only one of these clashes to have had such an outcome for patriarchy to be established" (Chevillard and Leconte 1986, 106–7).

The authors do not agree, however, on whether patrilocality and its development into societies characterized by male dominance occurred gradually over time or involved a violent confrontation and subjugation of women. Either way, lineage societies of various types have had an extremely long history, and still exist in various parts of the world (although they have all been partially transformed through colonialism and imperialism, usually in the direction of male dominance through the patriarchal ideology of the conquerors). Their full development is characterized as "kin corporate societies" to differentiate them from the kinship-based kingdoms or states which developed from them. These latter were in turn supplanted by civil states based on absolute property rights (112).

> The development of kin corporations out of communal society ... greatly increases the potential for social inequality, setting in motion processes and conflicts that may culminate in the establishment of a true class structure. These early class systems, which we call kinship-based kingdoms or states, retained many of the prior forms of kinship organization, even while social relations were transformed and class differences were institutionalized in the social, political, economic and religious spheres. (Coontz and Henderson 1986, 142)

The aristocratic kingdoms of Homeric Greece constitute an example of such a society. Coontz and Henderson (1986) argue that such kingdoms were characterized by the devaluation of *women as a category*, meaning that individual women, especially those in the ruling classes, could and did exercise considerable power and authority, although their power was "derived" from the men with whom they were linked through kinship, including marriage. "Women only rarely participated in relations of patron-clientage, being more commonly added to lineages as household slaves, unless they were acting in the capacity of men – that is, as *female husbands*" (144; emphasis added). Chevillard and Leconte (1985, 156) argue that the emergence of slavery was based on the subordination of women. Coontz and Henderson (1986, 145) note in this regard that conquered groups in archaic states were conceptualized as female "because their obligations to the patrilineal central powers were so like those of the wives within its lineages". And it was precisely because "women as a sex" were already excluded that individual elite women could gain access to power.

However, these ancient states were based along kinship and lineage lines. By contrast, the civil state imposed "an entirely new principle of association – territoriality – on its citizens" (Coontz and Henderson 1986, 150). The example the authors use here is the classical Athenian city-state. They note that the emergence of the civil state there was accompanied by the institutionalization (and the rationalization) of slavery. Most important, kinship was no longer to be the central principle for political organization and unification. The new ruling classes who brought the civil state into being were thus threatened by the old order and had to destroy its underlying relations. They did so by trying to exclude kinship as a basis for political participation, by creating a "private sphere of life which was separate from but subject to the public sphere organized by the state ... Where marriage once united corporate kin groups, it now broke them up, as husband and wife were urged – and often legally obliged – to identify with each other rather than with their kin. The state wished to deal with separate households, which it could control more easily than extended kin corporations" (150). Coontz and Henderson provide the development in Greece as an example. There the household (*oikos*) replaced the clan (*genus*) "as the integral, organic unity of the *polis*. To the *polis* instead of the *genus* the family *oikos* began to look for protection ... and to the *polis* in return the family *oikos* owed a variety of obligations" (151).

This newly stratified society also established patriarchy, since it was the head of the household who represented *his* family in the public sphere of *polis* life. Indeed, citizenship was categorically denied to all women. Women were confined to the private realm of household and made into the legal juniors of the household head, who, in turn, became a legal junior in relation to the state. "The increase in the male's authority over the nuclear household disguised his loss of authority over the extended family of labourers and reproducers" (Coontz and Henderson 1986, 153). Coontz and Henderson also note that women were associated with the old order in ancient literature, made to represent "chaos, civil disorder, and primeval forces" while men came to be associated with the virtues developed in the new civil state: "law, patriotism and rationality" (152). Saliou makes the further point that the establishment of "democracy" in Athens disempowered even individual privileged women who had previously had some access to power because it would have had to extend the "right to speak up" to all citizens' wives.

Saliou (1986, 192–3) notes that the status of women was higher in Sparta, which retained an aristocratic government. She provides another, more recent, example in the period after the French Revolution, when the establishment of the liberal democratic state was accompanied by women losing their political rights. Similarly, when the government of New France was transferred to England, women also lost political rights (the franchise based on property holdings) and economic rights to property which they had previously enjoyed. In 1849, following the Act of Union that joined Upper and Lower Canada, a law was passed that "specifically excluded women from the franchise in both of the Canadas" (Prentice, Bourne, et al. 1988, 99). In the previous chapters we have also seen that aboriginal peoples and Chinese, Japanese, and East Indian Canadians were also disenfranchised *after* the establishment of the liberal democratic state in Canada.

The liberal democratic state in western Europe thus emerges as the product of a complex series of struggles involving the subordination of women both as a category and as individuals, the establishment of slavery, and the establishment of class exploitation that involved the transformation of kinship to strip households of kin-based systems of power and to redirect these in a new form of allegiance to the ruling classes. The final product represents the full development of male dominance in patriarchy and class exploitation based on the extraction of surplus labour power from the individual, who is thus exploited in economic relations at the same time as he is granted rights of citizenship in the political (public) realm. Whole categories of people are initially excluded from the rights of citizenship, which translates into double exploitation of their labour power in the economic realm.

The private/public dichotomy also develops within the context of the civil state. In the private realm are confined not only women and their children but also slaves. Engels (1981, 121) notes that the very word "family" is derived from the Roman *familia*, whose original meaning was "the total number of slaves belonging to one man ... The term was invented by the Romans to denote a new social organism whose head ruled over wife and children and a number of slaves, and was invested under Roman paternal power with rights of life and death over them all."

The preceding analysis demonstrates the contradictory processes that culminated in the establishment of the liberal democratic state. Once established, however, these contradictions continue. The disenfranchised use the very notion of the rights of the citizen to fight for inclusion in the public realm, at the very least demanding the right to vote. Since the state is not directly controlled by any one dominant class (the Miliband/Poulantzas/Laclau debates are instructive here), individuals and groups can put pressure on politicians, law makers, and government bureaucrats to change the laws. Unfortunately, this does not result in equality, as Marx pointed out in his work. It gives the illusion of a true democracy while masking the inegalitarian economic basis that ultimately supports the concrete functioning of the state. The contradictions and struggles continue, but also shift from the advanced capitalist nation states to the underdeveloped countries of the world.

Simultaneously, the dichotomization of society into private and public realms remains but continuously shifts ground. The changing definition and relations between the private and the public is well illustrated by re-examining the political thought of Arendt, as in chapter 2. For example, in the late nineteenth century salmon canners did not have available to them the labour power of European women and children confined in the nuclear family household. The canners thus used the labour of aboriginal women and children, Chinese men, and Japanese women. All of these *categories* of labour power were also devalued, excluded from the public realm.

This analysis suggests the need to think far more creatively about the interplay between the private and the public realms in the capitalist mode of production. Feminist analysis has tended to treat the two as somehow permanently fixed and immutable, especially as evidenced in the debates occurring in the 1970s about the existence of a domestic mode of production within capitalism. For example, Stainsby has faulted me for ignoring women's double day of labour in my treatment of shoreworkers. While women's extra burdens in looking after children and husbands are certainly pertinent today, the very categorization of cheap wage labour in the nineteenth century was based on the fact that the reproduction of labour power occurred elsewhere, in China and Japan. Aboriginal women certainly looked after their children but often in direct combination with factory work since families lived on site and children worked alongside the adults. Criticism was later directed against Japanese women because they worked with their babies strapped to their backs. All these realities force us to acknowledge the artificiality of separating out the private from the public spheres, a dichotomization that is founded in patriarchal thinking. While Stainsby's analysis provides much needed information on the current mode of exploiting women's wage labour in the fisheries, it is important not to fall into a dichotomized mode of analysis that reproduces the categories that need to be "deconstructed." I would also question that the labour of women can be separated from issues of race and ethnicity in an industry such as this one. Split labour market theory tends to replicate the same type of dichotomized thinking by splitting labour forces into two spheres: primary and secondary. This is a description of a reality that is in need of deeper probing.

PROPERTY RIGHTS IN THE FISHERIES

The preceding analysis suggests a manner of theoretically reformulating an understanding of the liberal democratic state to incorporate the concept of cheap wage labour and the role of the emergent civil state in excluding categories of people from the rights of citizenship. Such a reformulation thus permits the conceptualization of the relations embodied in and through the state as in "relative autonomy" with regard to the other sets of relationships that develop within a capitalist mode of production.

The emergence of the nation state is thus also implicated in the valuation of labour and how it comes unto the labour market. We saw in previous chapters how the various states in Britain, China, Japan, the United States, and Canada entered into negotiations, as a result of the armed subjugation of China and Japan by the imperialist powers, over the movement and use of labour from China and Japan to the United States and Canada. While this movement of labour was connected to the emergence of capitalism in the Pacific Northwest and elsewhere, negotiations were not only or merely conducted to satisfy capitalist requirements for cheap labour power. For example, the merchant classes in China and Japan used these opportunities to consolidate their class interests both within their own countries and in those countries which saw the establishment of large settlements of Chinese and Japanese residents.

While the use of labour power within emerging capitalist relations of production leads to contradictions and conflicts which both facilitate and hinder the interests of the capitalist classes, property rights also undergo transformation. For example, the emergence of private property rights is connected to patriarchy and the dichotomization between the private household and the public realm. The fisheries of

British Columbia have also posed unique problems surrounding the issue of property rights to a fugacious resource. The federal state from the outset regulated the fisheries in such a way that both canners and fishers complained about over-regulation, especially in comparison to the relatively lenient administration of the U.S. fisheries in Puget Sound and Alaska.

Much of the literature on the fisheries treats fish as "common property" until caught, at which time they become the private property of those who catch them (Marchak 1987, 4). Marchak argues that the fisheries should instead be conceptualized as "mismanaged state property" (5). For many analysts use the "fact" that the fisheries are common property to explain the depletion of stocks and the erosion of the fisheries since the end of the nineteenth century. Marchak also uses the term "communal property" to characterize the use aboriginal peoples made of the fisheries and the way in which they collectively organized to conserve and manage the resources they relied on for subsistence. She notes that the difference between the two forms of property in the fisheries is that between using the resource for subsistence activities and selling it for profit – the commodification of fish. "Once accumulation, rather than subsistence, is the reason for catching fish (or cutting trees or any other activity), there is a need to define and defend property rights; without such definition, individuals and companies would be unable to ensure that they, rather than any others, should benefit from their investments and activities. One property holder could defend his or her property by force, as separate fiefdoms have done in earlier times. But a superior method of defining territory is a system of laws ultimately backed up by a single institution with a monopoly on force to which all members of the territorial unit are equally subject. This institution is the state" (Marchak 1987, 11).

Marchak also notes that these property rights preceded the establishment of the state. In other words, the state became necessary to regulate the conflicts generated by competing claims. Thus governments come to regulate not the rights of individuals and specific pieces of property but rather the entire system of property rights and the corresponding system of accumulation. The provincial fisheries are particularly interesting in this context both because the resource is not fixed (moving in waters that lie within and beyond the jurisdiction of the Canadian state) and because big capital in the form of the major processing companies competes for the resource with aboriginal peoples concerned with their food fisheries, with small processors, and with dependent commodity producers or fishers.

Thus the federal state has been centrally involved in the conflicts between processors and fishers because it controls the "management" of the resource. Ever since the federal Department of Marine and

Fisheries was created in 1867 and the Fisheries Act was passed the following year, the federal state has had a strong presence, first on the Atlantic seaboard and subsequently on the Pacific. It has a long history of legislation, including licensing boats on the west coast in the hope of restricting the numbers of fishers and thereby limiting the capture of fish to preserve stocks (with very limited and dubious success). Other regulations have included restrictions on type of gear (for example, banning traps with one exception), length of the season, and times (some fisheries are opened for a few minutes, during which the catch allocations are met, indicating the high levels of technological development in catching capacity and capability). McMullan (1987, 107) has outlined three policy periods: an early regulatory phase in which the state created property rights, a modernization stage associated with the creation of wartime advisory boards, and the period since 1968 with renewed licensing, regulation, and control.

Thus the federal state has had a direct impact on both processors and fishers, enabling both to continue operating in the long run while severely curtailing the operations of either or both in the short term. But the state has also been subject to direct and indirect pressure from both sets of interests, causing it to modify its laws and their enactments. It has also been subject to pressure and negotiation on the international stage. In the nineteenth century, control of the Atlantic fisheries almost caused war between Canada and the United States. More recently the United States has invoked the GATT agreement to force Canada to sell unprocessed fish across the border, and free trade has strengthened the U.S. argument for unregulated flow of raw, unprocessed fish across the border. This threatens processing companies and jobs for shoreworkers on both the Atlantic and Pacific coasts. The strength of the UFAWU has lain in its ability to pressure the federal government to ban the sale of unprocessed fish where it can be processed more cheaply by underpaid labour in Alaska, Mexico, etc., thereby protecting the wages of shoreworkers who legitimate the organization's claim to be an industrial union.

Since previous chapters have focused on the early history of the industry, a brief summary of the state's more direct involvement in capital formation and in labour relations in the early period will serve to clarify and extend the analysis.

STATE INVOLVEMENT IN CAPITAL FORMATION

As mentioned, the federal state came to play a central role in the fishing industry upon Confederation, when it assumed responsibility for the management of the fisheries. But capital formation in the nascent salmon canning industry reflected older colonial ties, linking what became British Columbia in a triangular set of relations with the merchants of San Francisco and with British markets by way of merchant capital centred in Victoria (Ralston 1981). The federal Canadian state reflected a new pattern of capital formation, linked to the western movement of banking capital and the building of a federally subsidized transcontinental railroad.

Salmon canners initially developed ties with local politicians and merchants. And they formed organizations like the Salmon Canners' Operating Committee to represent their interests at the political level. The various fisheries reports also provide evidence that salmon canners participated directly, sitting on the boards of various trade councils, such as the one in New Westminster (the original urban centre of the Fraser River fishery). These councils in turn carried weight with the provincial state. The connections between capitalist and legislator were often direct. Thus McDonald (1981, 374–5) notes that, until 1898, political ideals were based on "the exploitation of regional resources through the provision of public concessions to private interests. Victoria businessmen constituted both the governments that granted and the interests that received these concessions. While Victoria merchants controlled hinterland trade and development, Victoria government promoted hinterland development."

As noted in chapter 3 the westward movement of the banks served to displace the old mercantile interests centred in Victoria and enabled salmon canners to assume more direct control over capital financing. However, these changes also entailed the loss of mercantile domination in the provincial economy of the salmon canning industry. McDonald notes that the industry represented the traditional maritime connections that characterized the nineteenth century. But in the twentieth century, railroads came to displace ocean transport as the primary means of moving commodities, and established new markets which salmon canners were unable to develop. While U.S. canners found a market in the southern states, the Canadian population was too small to provide Canadian processors with a domestic market. But another, equally important factor was the lack of state aid in subsidizing rail transport to help lower costs and thereby facilitate the development of such a domestic market.

Victoria underwent deindustrialization. The value of manufacturing fell by half. The immediate cause was the depression of the early 1890s, which hit the city particularly hard. Over the long run local iron works, shipbuilders, and other companies were simply unable to compete with central Canadian enterprises. Ontario manufacturers, with their large population centres, could produce on a much larger scale, and they did not have to contend with the CPR's discriminatory rate structure. For instance, it cost more to send manufactures from west to east than vice-versa. (Barman 1991, 111)

The emergence of the Canadian federal state was closely connected to the building of a transcontinental railroad and to the financial interests connected with the project. The connections among the first post-Confederation cabinets, the financial bourgeoisie, and the railway entrepreneurs were quite close: "They were often the same people" (Panitch 1977, 11). In financing the railway the federal state thus came to be directly involved in capital accumulation. Dependence on British financial backing also continued. And in order to meet the debts incurred during construction, the federal state began to encourage immigrant settlement of the vast prairie the railroad had to traverse (Fowke 1957). The result was uneven regional development. Canadian industries in central Canada sought to develop western markets for their commodities, while western producers were encouraged to ship by rail unprocessed or semi-processed natural resources, like wheat and lumber, to eastern Canada. Federal state regulations encouraged this two-way flow by instituting tariffs and subsidies by type of good transported.

In addition, the development of the U.S. economy also encouraged, and redirected, regional dependency. After the First World War, Canadian resources were increasingly in demand to fuel U.S. industry. Most natural resources came under the jurisdiction of the provinces, and thus regional interests began to make alliances with the United States rather than with the federal Canadian state (Stevenson 1977, 78). However, like central Canadian industry, Americans wanted resources in an unprocessed or semi-processed stage, establishing Canada as a market for these same resources in their final manufactured form. Thus, Panitch among others argues that the Canadian state in this regard has been marked by a *lack* of relative autonomy *vis-à-vis* the dominant class interests. "It has been the very lack of relative autonomy of the state, the sheer depth of its commitment to private capital as the motor force of the society, which, when combined with a weak indigenous industrial bourgeoisie and a strong financial bourgeoisie cast in the mould of an intermediary between staple production in Canada and industrial empires abroad, explains the length to which the state has gone in promoting private capital accumulation not only for the domestic bourgeoisie but for foreign capitalists as well" (Panitch 1977, 16–17).

One of the casualties of these new developments was the B.C. salmon canning industry. Upon completion of the railroad, tariffs were put in place to favour the movement of certain types of goods in specific directions. The result was that U.S. traffic was favoured over Canadian, eastern over western, the movement of raw materials east and finished goods west, and long-distance over local traffic (Naylor 1975, vol. 2, 26). In British Columbia these trade patterns helped stimulate the movement of timber and minerals in an unprocessed or semi-processed condition out of the province. No attempts were made to help promote domestic fish consumption until 1908, when the federal state undertook to pay one-third of the expenses on less-than-carload lots of fresh/frozen fish from the Pacific to the eastern boundary of Manitoba (the Atlantic provinces received the same subsidy for shipments to eastern and central Canadian markets). Thus while the Canadian federal state may be seen as lacking relative autonomy in its direct connection with capitalist class interests, that these interests are also divided and unevenly represented (Poulantzas' [1978, 77–85] concept of class "fractions" is appropriate in this context) is demonstrated by the manner in which salmon canners were disadvantaged in the new alignment of federal and provincial state powers. Fisheries officers also recognized the anomaly: "Fish is a home product. It costs nothing to cultivate, and the capital invested in the fisheries in comparison with the yield, is smaller than in any other food producing industry. Fish should, consequently, be a cheap food in all parts of the country, but to make it so, adequate transportation facilities at moderate prices, must be available" (DMF 1917, xv).

Because fisheries officers tended to view the provincial salmon canning industry as over-developed in comparison to other fisheries, the state made no attempts to develop domestic markets for canned salmon. Salmon canners continued to rely on foreign markets in which other countries dominated. And the federal state virtually handed the halibut fishery to U.S. interests. The New England Fishing Company, which established a base in Vancouver in 1894, was given special privileges. Beginning in 1897, U.S. fishing vessels coming to B.C. ports were allowed to ship their fish in bond to the United States (*SP*, no. 39, 1919, 8). In 1915, after completion in 1914 of the Grand Trunk Pacific railway to Prince Rupert, this privilege was extended. Fish caught in U.S. bottoms could be shipped to the United States over Canadian rails. The intent of state authorities was to develop Prince Rupert as a railway terminus, port, and centre of the halibut fishery by encouraging buyers to locate their business there. The result was U.S. domination of the Canadian halibut fishery.

Problems also occurred over the lack of a duty on fresh fish (other than sockeye salmon) shipped to the United States. American canners, unlike their B.C. counterparts, had a domestic market for the cheaper grades of canned salmon, a market protected from Canadian competition. Thus U.S. canners could afford to pay Canadian fishers higher prices for these species of salmon. While fishers benefited, Canadian canners were deprived of both the market and the opportunity to process this fish. The inability of the federal state to negotiate the removal of the U.S. duty on Canadian canned salmon, depriving Canadian processors of the large American market, illustrates the extent of Canadian dependency on the U.S. economy (*SP*, no. 39, 1918, 231).

These examples indicate the contradictions inherent in the state's approach to the fisheries of British Columbia. Also implicated are the U.S. levels of government. On the Atlantic seaboard the two countries almost went to war over control of the fisheries. And on the Pacific, the salmon fisheries on Puget Sound and on the Fraser River have been a constant source of friction. With the exception of J.H. Todd, who received special political dispensation off Vancouver Island, traps were allowed in U.S. but not in Canadian waters. Canadian processors and fishers often complained that the much more stringent regulations in Canada were in effect preserving the resource for the U.S. industry. The U.S. government continuously blocked a treaty regulating salmon catches on Puget Sound and the Fraser River, and it was many decades before such a treaty finally went into effect.

When we investigate the intricacies of capital accumulation in a particular industry like the B.C. fisheries, the complexity of the various competing interests, including those of government officials and of politicians, becomes striking. In fact the history of capital accumulation in this industry and the role played by the Canadian and U.S. states is vastly more intricate and complicated than this short summary has allowed. When we turn to the role of the state in labour relations, another brief summary must be given of the extremely racist ideology the provincial state helped to nourish among the white population in British Columbia. It serves to show yet another side of state relations. In establishing its own legitimacy and hegemony, state officials use embedded visions of the world to promote their own importance in governing without continuous recourse to armed force.

RACISM AND THE B.C. STATE

Because the nation state emerges as a response to contradictory sets of interests that compete for power, it is connected to the creation and dissemination of an ideology that serves to unite classes otherwise opposed to one another at the economic level. The ideology of liberal democracy, the creation of citizens as political equals, is one example.

From the mid-nineteenth to the mid-twentieth century, an ideology of white supremacy also united British Columbians against perceived economic, cultural, and military threats from non-British peoples.

When it was founded in 1871, British Columbia was composed of three distinct groups: Europeans, Chinese, and aboriginal peoples. By the end of the century, the aboriginal population was so decimated that European residents no longer perceived it as a threat.

> That a deep and permanent racial cleavage divided British Columbia its white residents were well aware. In assessing the origins of this racial division, however, they could largely ignore the native Indian population for it had long since been pushed aside and seemed to languish in decay. The Asian community, on the other hand, was a dynamic, growing segment in west coast society. Its continued expansion promised only to broaden the racial fissure which already fragmented this society.
>
> Cultural pluralism, then, was unacceptable to the white community. Within it the plural condition generated profound, irrational racial fears. Pluralism stirred a deep longing for the social cohesion which could only be achieved, it seemed, by attaining racial homogeneity. (Ward 1978, 92–3)

Ward offers a psychological explanation for the racial ideology that emerged in British Columbia. "For the most part those who have previously studied the Asian question in British Columbia have assumed, if not concluded, that anti-Orientalism was grounded in economic tensions created by the availability of cheap Asian labour in a maturing industrial capitalist economy. My argument, on the contrary, holds that racism in British Columbia was fundamentally a problem in the social psychology of race relations" (Ward 1978, ix). While many of his findings will be explored in this section, the theoretical analysis developed in this and earlier chapters suggests that racism is a social construction rooted in complex and specific sets of social, cultural, economic, and political relations and that patriarchal consciousness also informs the ideology that categorizes groups of people by the colour of their skin.

In his 1987 study of anti-Semitism in Canada, Stanley Barrett offers a definition that parallels the analysis developed here.

> Racism, particularly in the context of black-white relationships, is often portrayed as universal and natural, not just by white supremacists, but also by some scholars. Yet that is not true ... Racism constitutes an elaborate and systematic ideology; it acts as a conceptual tool to rationalize the division of the world's population into the privileged and the deprived. It is inherently a political phenomenon. Racism in this sense is a relatively recent thing ... Most

specialists agree that it emerged with the advent of the colonization of the Third World by European nations, and thus coincided too with the development of capitalism. (5–6)

Barrett goes on to caution against treating racism in an "economistic" manner, as capable of being explained solely in class terms. "No major social phenomenon is unidimensional. Not only has racism in the past century ventured beyond its economic cage, so that today it possesses a semi-autonomous political dimension, but it has also become criss-crossed by a range of factors including deep-seated psychological ones" (S. Barrett 1987, 7). The racist legacy of British Columbia is particularly instructive because it shows the implication of the state in perpetuating a racist consciousness that ran counter to the interests of at least a portion of the capitalist class. However, fanning racism within the white working class and among dependent commodity producers (especially fishers) did serve to divide groups of workers against one another, thus retarding the formation of class consciousness that would unify labourers in their relations with employers.

Barman (1991, 134) refers to the development of "scientific racism" following the publication, in 1859, of Charles Darwin's *Origin of Species*, which was quickly "perverted into a concept legitimating racism based on biological characteristics. Out of the inevitable struggle to exist, so the adherents of social darwinism argued, the fittest survived, be they animals or human beings. To determine which of the races was the fittest, it was only necessary to compare western technological achievements and the growth of capitalism with the rest of the world." The term "survival of the fittest" was in fact coined by the sociologist Herbert Spencer (Hale 1990, 141). Frank Burnett, president of United Canners, Henry O. Bell-Irving (A.B.C. Packers) and pioneer salmon canner Alexander Ewen all used the phrase in their testimony to the Royal Commission of 1902 (138, 144, 150). Barman (1991, 141–2) pushes the argument even further:

The racism that legalized discrimination against Asians was extended, in varying degrees, to all arrivals. In the sense that British Columbians of the dominant society were racist – and indeed they were – they did not restrict their antipathies and prejudices. Visible minorities were but the tip of the iceberg. It was the almost universal assumption, held not just in British Columbia but across English Canada, that immigrants should learn to conform to an Anglo-Protestant stereotype which those in positions of social and economic power equated with themselves and with them to the prestige and pre-eminence of the British empire.

From 1858 until the exclusion act of 1923, Chinese immigration witnessed significant peaks. British Columbia in the middle of the nineteenth century was a gold-mining frontier, marked by gold rushes during which the population swelled only to ebb once the gold fever was spent. In this period, the Chinese formed a significant proportion of the population: over 20 percent by the early 1880s. Large numbers of Chinese labourers were brought in to construct the railroad. However, by 1921 European immigration far outnumbered Chinese who formed less than 6 percent of the provincial population (Ward 1978, 15).

Attitudes against the Chinese began to harden in the 1860s, when an economic depression left many miners unemployed. The first attempts to discriminate legally against them were not made, however, until after Confederation. Provincial politicians passed a bill to exclude the Chinese from the provincial franchise. In 1878 the Chinese Tax Act was passed, requiring all Chinese residents over twelve years of age to pay a quarterly tax of $10. Salmon canners protested the tax. And Chinese labourers in Victoria went on a strike that lasted several days, refusing to work for white employers. The B.C. Supreme Court declared the act *ultra vires* provincial jurisdiction.

Employers who required large Chinese labour forces found themselves in conflict with provincial politicians. At the Royal Commission hearings in 1885 and 1902, salmon canners testified that the industry would not have succeeded without Chinese labourers. Although they received wages one-third to one-half lower than those paid to white male labourers, they had developed important skills in certain industries. "The Chinese cannery worker at the turn of the century was typical ... He possessed significant industrial skills, a fact acknowledged by his employer, and he stood on a middle rung in the province's labour hierarchy" (Ward 1978, 16). But immigration restrictions reduced the available number of Chinese labourers, and those who remained could command higher wages. Salmon canners responded by mechanizing those parts of the labour process Chinese labourers controlled. It is significant that the "Iron Chink" was adopted after 1905, the year head taxes were raised.

Although the provincial state continued to enact restrictive legislation, it was blocked at the federal level. Large numbers of cheap labourers were required to build the railroad through the Rockies. In 1882, Onderdonk informed Prime Minister Macdonald that Chinese labour was essential if the railroad was to be completed within a reasonable time. Not only were the restrictions the B.C. government sought to enforce overturned, but the numbers of Chinese labourers entering Canada increased significantly. Even after the railroad was finished, the interests of the CPR continued to run counter to policies favouring restriction. Its steamship line had developed a lucrative transpacific passenger trade, and immigration quotas threatened to cut into its profits (Ward 1978, 35, 59).

The federal state, while often forced to nullify provincial legislation, also developed an ideology of exclusion that was racist in intent. Macdonald had to mollify interests connected with the railway and thus could not, for the moment, restrict immigration. Upon pressure from eastern labour unions, he did, however, introduce an amendment to the Franchise Act, preventing any "person of Mongolian or Chinese race" from voting in federal elections. The term "person" in the Franchise Act was "not intended to include Chinese or Mongolians" (Ward 1978, 40 and 179n.17). From 1885 until 1947, the year in which Chinese and East Indian Canadians were granted the federal and provincial franchises, the state considered neither Chinese residents nor status Indians (as defined by the Indian Act) to be persons. Roy (1989, 63) notes that as early as 1882 Macdonald referred to the Chinese as "an inferior race," as "semi-barbarians" and as "machines .... In short, John A. Macdonald was a racist." Nor was Prime Minister Sir Wilfred Laurier much better: "He believed that it did 'not matter very much' how the Chinese were treated since 'they submit to contempt and kicking.' Indeed, he told a Montreal reporter that while the Japanese were akin psychologically to the Chinese, they differed 'from them as the fully developed man from the child' because the Japanese had adopted western ideals of government" (203).

In the 1890s Japanese immigrants began to enter Canada in significant numbers. Provincial politicians extended legislation to incorporate new groups. While the Japanese were disenfranchised in 1895, in 1907 East Indians were excluded from the right to vote. Naturalized and Canadian-born residents in these groups were also deprived of the right to vote.

In a series of "Gentlemen's Agreements" Japan agreed to voluntarily restrict emigration to certain countries. In 1902 Japan signed a diplomatic alliance with Great Britain, and emigration restrictions were extended to Canada. For this reason the Japanese were originally considered to be a lesser threat than the Chinese. However, the restrictions were lifted in 1907, the year significant numbers of Japanese immigrants arrived via Hawaii. In addition, since 1903 the numbers of Chinese immigrants had been rising. Several emigration agencies in Vancouver negotiated contracts with large Canadian corporations to provide them with Japanese labour. In 1905 Japan defeated Russia and came to be perceived as a military threat. All these events were factors that led to the riot in 1907 when white residents stormed the Chinese and Japanese communities in Vancouver, destroying property. "The riot placed the Laurier government in a rather awkward position. It was forced to placate both Japan and British Columbia simultaneously" (Ward 1978, 73). Japan agreed to voluntarily restrict emigration to Canada to 400 per year.

But racial issues were not paramount solely in British Columbia. They formed part of the platform of both the federal Liberal and Conservative parties in their 1908 election campaigns. Future Prime Minister Borden pledged that the Conservative Party "will ever maintain one supreme consideration to which all material considerations must give way; and it is this: British Columbia must remain a British and Canadian province, inhabited and dominated by men in whose veins runs the blood of those great pioneering races which built up and developed not only Western, but Eastern Canada" (Ward 1978, 75).

In 1908 the federal government approved an order in council that effectively closed the door to further Japanese and East Indian immigrants. All immigrants were prohibited from entering Canada unless they came from their own country of birth or citizenship by a "continuous journey and on through tickets" purchased in their home country. There was no direct steamship route from either Japan or India. "The order provided effective restriction while avoiding the distasteful and increasingly unacceptable practice of indicating undesirable immigrants by race or nationality. Henceforth this order and the Lemieux agreement, together with the Chinese Immigration Act, formed the new foundations of the Liberal government's Oriental immigration restriction policy" (76).

The East Indian community successfully challenged the legality of federal immigration restrictions, resulting, in 1914, in the *Komagata Maru* incident. The ship's passengers were prevented from disembarking in Vancouver. "When compared with the numbers of immigrants who annually entered the province, the 376 on board the *Komagata Maru* in 1914 were of very little significance. Furthermore, aside from a handful of previous residents, none of the prospective immigrants landed. Yet while the ship lingered in port, racial tensions remained high, as high, indeed, as they had been in 1887 and 1907" (Ward 1978, 92). A local newspaper proclaimed: "There are 300,000,000 natives of India behind them, who have the same rights as these" (Ward 1978, 90). Ward notes that most of these immigrants, like Chinese and Japanese immigrants at that time, were of peasant background. They also left their families behind, hoping to earn enough in Canada to ease the economic plight of those left behind. Very few, if any, East Indian men were employed in the fisheries. However, when they later began to bring their families to Canada, East Indian women found employment in the fish plants. (While in Prince Rupert in May and June, 1982, as part of the Fish and Ships research team, I spoke to some of the East Indian families living there. I interviewed a man and his father who came from Punjab in India. The son, who was 23 years of age, noted that some of the East Indian women had been working in the Prince Rupert Fishermen's Co-operative for ten years. One woman had been there fourteen years. Most of the Punjabi emigrants had farming backgrounds. The son told me he was involved in an organization fighting the Ku Klux Klan and he estimated that racism in Prince Rupert was high. He expressed a desire to return to India. He had eight brothers and sisters. It was his sister who sponsored their parents and two of her brothers. We also spoke to several men who had been managers of canneries. For example, one had managed canneries for B.C. Packers from Steveston on the Fraser River all the way up the coast [Namu, Sunnyside, Port Edward, etc.] to Alaska. A number of these men had entered fishing as teen-agers, following in their fathers' footsteps. One former plant manager told us that when he worked at Imperial cannery in Steveston, they used to hire East Indian women when they needed lots of extra hands. He referred to an East Indian "contractor" who demanded that the women's pay cheques [ten women were involved] be given over to him. When the manager refused, the man waited outside the cannery and as the women filed past they handed their cheques to him. A number of industry people made references to a continuation of the contract system using this ethnic group as evidence of the continuing prevalence of the contract system in the industry. Although he doesn't examine this particular ethnic work force, Neil Guppy [1987, 199–222] provides data on the relationship between ethnicity, wages, and breakdown of work tasks stemming from the Fish and Ships research findings.)

From the early 1890s to the 1920s, Japanese men came to represent a significant proportion of the province's commercial fishers. Initially, immigration was confined, as it was for the Chinese, to unmarried young men. But between 1910 and 1930 the situation began to change. While in 1911 the ratio of Japanese men to women was about 5:1, by 1921 it was 2:1 and decreased further in the following years. In comparison, the proportion of Chinese men to women in 1911 was 28:1, and by 1921 had decreased only to 15:1 (Ward 1978, 109, 187). Japanese women, who began to arrive in British Columbia during the period when fresh/frozen fish markets and refrigeration techniques were being developed, took many of the new jobs these areas opened, predominantly on the assembly lines filleting fish.

As the proportion of Japanese to white and aboriginal fishers rose, the federal state was pressured to limit their numbers. It eventually adopted a policy to eliminate the Japanese altogether from the provincial fisheries. In addition to becoming commercial fishers, these immigrants also developed a lucrative business dry salting fish for export to Asian markets. Federal fisheries officers developed policies aimed at placing control of this industry in the hands of white or aboriginal entrepreneurs. "In view of the policy of the department looking to the elimination of the Orientals the industry should be totally in the hands of the white population and Canadian Indians by 1927" (DMF 1924, 54). The Department of Marine and Fisheries did not have to develop a similar policy for the Chinese since provincial legislation had blocked immigration and their numbers had declined.

In the salmon canning industry, the state was notable in its absence. While the federal state acquired management of the fisheries, the provincial government was responsible for the canneries (although which level was to receive licencing fees remained a bone of contention for several years). In the 1920s, the provincial government began to use labour legislation to restrict the employment of "Asian" labourers. "The minimum wage law enacted in 1926 was intended to enforce a basic wage in selected industries; if this were done it would eliminate that competitive edge which Orientals enjoyed" (Ward 1978, 137). This was not necessary in the salmon canning industry, however. White male labourers never expressed much interest in doing the work assigned to Chinese and aboriginal women. They found employment in other areas of shore work, as machinemen and in the reduction plants (a highly mechanized part of the industry), and were taken directly onto company payrolls. The seasonal labour forces, in contrast, continued to be employed on contract and thus did not compete with permanently employed staff. Urban growth and an influx of non-English-speaking immigrant women provided an alternative source of labour for seasonal work to Chinese male labour power.

Until the 1930s, the Canadian labour movement was also generally racist in its outlook and organization. The situation began to change during the Depression, but in the fishing industry there was a last surge during the Second World War when Japanese Canadians were forcibly evicted from coastal British Columbia. Many fishers actively called for their removal and benefited materially by acquiring the fishing vessels of Japanese fishers at nominal prices.

Working-class pressure during the Depression and war years forced changes in labour legislation. Trade unions and the right of collective bargaining were legally recognized. Fishers and trade unionists affiliated with the Communist Party organized an industrial trade union to cover the entire provincial fishing industry. Minimum wage legislation was enacted and extended to cover most industries, rendering inoperative systems like the Chinese contract. As part of the new legislation, employers had to list their employees by name on company payrolls, thus eliminating the advantages employers enjoyed in contracting labour out to middlemen. Along with labour legislation that led to improved wages and conditions of work, groups began to press for an end to discriminatory laws and practices. Nevertheless, immigration laws continued to be phrased in ways that selected which groups and individuals were allowed entry into Canada. In 1947 the federal state repealed the Chinese Immigration Act, although immigration from China continued to be restricted. That same year Chinese and East Indian Canadians were granted the federal and provincial franchises, and these were extended in 1949 to include Japanese Canadians.

The franchise is a crucial point of entry to participation not only in political but also in economic life. Macpherson (1979, 47) argues that democracy is redefined within capitalism: "The old idea of democracy, as rule by and for the poor, had been converted to the idea of democracy as the right to get into the competition." The state influences relations of power, and in a society where access to the means of labour is blocked, political and economic power become interrelated. "So in any society where the legal institutions give all the property in land, or any other means of labour, to one section of the people, all the others must pay for access to the means of labour ... Whatever form the payment takes, it is a transfer of a man's [*sic*] powers (or part of the produce of these powers) to another man, and it is compulsive" (41).

To enter into these power relations, even in a relation of dependency, as a seller of labour power, requires recognition by the state that one is an individual (a legal "person") capable of entering into a contractual negotiation. When the state refuses to recognize whole categories of labourers on the basis of their race or gender, then these groups are powerless politically as well as economically. While British male labourers bargained with employers by means of their rights as Canadian citizens to be represented by and through the state, these other groups remained peripheral to capitalist relations of production because they had no recourse to state intervention. When they eventually achieved political recognition as "persons" by acquiring the right to vote, they could then join organized workers to press for higher wages and better working conditions. But this eroded the capability of employers to hire cheap wage labour. The state stepped into the breach, developing social services and programs to maintain labourers as pre-capitalist relations of production were increasingly eroded and employers remained unwilling to provide in wages and benefits the full costs of producing and reproducing labour power. The development of the welfare state not only permits capitalist employers to pay wages below these costs, it also permits pre-capitalist relations, like those incorporated in the nuclear family household and on Indian reserves, to continue even when they can no longer fully supplement wages by providing unpaid work to maintain and reproduce labour power.

LABOUR AND THE WELFARE STATE

"The secret of capitalism is that *nobody* plans it" (I. Gough 1979, 29; emphasis in original). The relations that develop between classes are formed on the basis of the intervention by non-labourers in the ability labourers have of controlling their means of subsistence. Although there is an order and logic to these relations, there is no conscious plan. Neither can the state plan the economy, contrary to the claims of elected politicians. The state is itself a set of relations that develops partially from the economic contradictions between labourers and non-labourers. Those who exercise power in the name of the state can only hope to channel relations in a certain direction, one that will ensure the long-term interests of the capitalist class as a whole. Their task is made all the more difficult because, although relations between capital and labour develop logically, capitalism itself is illogical. The capitalist mode of production is rooted in the continual expansion of the forces and means of production, in the continual creation of surplus value convertible into profit.

> First, exploitation takes place automatically within the economic system; that is, the extraction of surplus labour does not require the political coercion, open or latent, of feudalism or slavery. The generation of surplus value is secured without conscious control by means of the market. Second, and due to this, the capitalist economy has a momentum or dynamic of its own which is again basically outside the control of any agent or class. Together these indicate that under capitalism the "economy" becomes separated from politics, the "private" sphere from the "public." The notion of a distinct political sphere is, therefore, peculiar to capitalism. The very individualism of capitalism, the fact that all subjects are formally free and equal to pursue their own ends, requires a separate structure, the state, to represent their "common interest." What results are the separate institutions of the modern state and their apparent autonomy from the relations of exploitation. (I. Gough 1979, 39–40)

The "paradox is that the modern state expressing a 'common will' ... only appears with the anarchic unplanned system of capitalism" (I. Gough 1979, 175). All these factors combine to make possible a number of political decisions at any one time, all constrained by structural circumstances. Class struggle at the political level also becomes possible, even desirable, since the state can then act to diffuse class conflict. The emergence of the welfare state within advanced capitalism can be understood as simultaneously a product of working-class struggle and a means of containing that struggle, of acknowledging it and allowing it expression within the system.

But the welfare state also emerges because two contradictions become acute: structured unemployment and the continuing production and reproduction of labour power within family formations. Ian Gough (1979, 44–5) defines the welfare state as "*the use of state power to modify the reproduction of labour power and to maintain the non-working population in capitalist societies*" (emphasis in original). The development of capitalism results in increasing strains in those areas of life not directly embraced by the market economy. The state helps to alleviate some of these strains, while at the same time it commodifies services previously undertaken in the non-wage sector. "For our purposes we may assume that the proportion of total use values provided by the welfare state is increasing over time by comparison with the capitalist and domestic sectors" (182).

The emergence of the welfare state in the post-war period had important consequences on the composition of labour forces in shore plants. Legal recognition of trade unions and the principles of collective bargaining displaced the services previously provided by contractors. At the same time, by replicating some of the services previously provided within the household, the state partially released the labour power of women. Coupled with developing technology that transformed the household and housework, women were released to labour for wages. In the fishing industry, expansion into new fisheries and the adoption of new processing techniques created additional assembly line jobs that became designated "women's work" (for example, pulling herring roe and canning crab).

Technological developments within the industry enabled fish companies to close remote plants and consolidate facilities in or near urban centres. Urban growth also provided them with a cheap pool of female labour power. Aboriginal women were forced to relocate to the cities or lose employment. As more and more plants in the central and northern areas closed in the 1950s and 1960s, aboriginal communities lost an important source of income and became ever more dependent on state aid.

Before the 1930s social legislation in Canada was minimal (Finkel 1977, 346). While the federal government was able to restrict trade union activity during the Depression, it was unable to stem it during the war years. Trade unionists also changed their tactics and began to organize mass-production industries, many under state control for the war effort. Membership rolls swelled as industrial workers joined the ranks of skilled craft workers. By 1944 Prime Minister Mackenzie King "could argue that class conflict had reached such a point that social legislation was the only alternative to socialism" (361). The organized labour movement demanded participation at the political as well as at the economic level. "Collective bargaining was not just a means of raising wages and improving working conditions. It was a demand by organized workers for a new status, and the right to participate in decision making both in industry and government ... Eventually this demand for a new status in society was met by the introduction of a new legislative framework for collective bargaining which has been modified only slightly since that time" (MacDowell 1978, 175).

While craft workers had always been able to negotiate on the basis of their skills, many mass-production labourers did not possess skills employers acknowledged as valuable. This made legal recognition of trade unions crucial. The fact that the federal government had guaranteed production in many of these industries, including the B.C. fisheries, for the war effort caused the state to intervene directly. While it agreed to hold wages at fixed levels, it was less willing to negotiate with labour. This led to a number of strikes in key industries in Ontario (such as steel) and the election of the CCF as the official opposition in that province. The federal government capitulated, and in February 1944 made trade unions legal. MacDowell (1978, 194) notes that it became a model for post-war legislation: "It guaranteed the right to organize and bargain collectively, establishing a procedure for the certification and compulsory recognition of trade unions with majority support, recognized the exclusive bargaining agency principle, defined unfair labour practices, provided for remedies, and outlawed company unions."

The B.C. fishing industry participated in the heightened labour unrest during the 1930s and the war years. Fishers recognized that craft organizations based on the type of gear they used in fishing divided them in their negotiations with the fishing companies. By contrast, these companies were unified in the Salmon Canners' Operating Committee which negotiated on behalf of most of the canneries in the industry. In the early 1940s fishers created an industrial union, and they began to recruit shoreworkers in addition to fishers. In 1945 union organizers used the new legislation to create the United Fishermen and Allied Workers Union, and for the first time labour legislation was systematically applied in salmon canneries. "Throughout the period from 1907 to 1934, the Province clearly had the constitutional authority to regulate labour in the fish-processing and fish-canning sectors; yet the application of the various Factory Acts, Hours of Work Acts, and Minimum Wage Acts all explicitly excluded protection and regulation in the fish canning sector, and the Workmen's Compensation Act seems to have been applied such that Asian and Indian labour had unequal access" (Garrod 1984, 17).

Legislation by itself is useless. It must be translated into regulations and enforced. The provincial state continually enacted racist legislation in the post-Confederation period. But much of it was blocked at the federal level, and was challenged by the groups against whom it was directed. While Chinese, Japanese, and East Indian Canadians had no voice politically, they could still negotiate with the federal state by putting pressure on diplomats and consuls from China, Japan, and India. Canada had to maintain friendly relations with these states and was thus forced to make concessions. And while labour legislation was drafted, it was not enacted for most shoreworkers as long as the Chinese contract system operated. In order to draft the regulations and make them effective, shoreworkers had to enter the fray and fight to be included in union negotiations, as the previous chapter demonstrated. However, as some groups attain parity with organized workers, employers continue to put pressure on the state to allow them to hire cheap wage labour. If this is no longer possible within Canada, corporations become multinational and seek to transplant their operations into countries where they can hire cheap wage labour. Certainly the neoconservative government of Brian Mulroney and the free trade agreements put in place with the United States and Mexico allow multinationals to do just this sort of thing and undermine trade unions like the UFAWU. The case study used here thus has applications to industries making use of cheap wage labour anywhere around the globe. This also suggests that the nation state faces severe constraints given the international/global movement of capital.

7 Conclusion: Marx's Labour Theory of Value Reconsidered

The theoretical starting point in situating and understanding the "making" of shoreworkers (Thompson 1979) in the industrialization of the B.C. fisheries has been Marx's labour theory of value. The basic problems with Marx's formulation as it applies to this particular case study are twofold.

First, Marx conceptualized the capitalist economy at its most abstract in order to uncover its basic "laws of motion"; that is, he postulated a capitalist mode of production in its pristine form. His purpose was to demonstrate how capitalism contained its own seeds of destruction, thus enabling its transformation through revolutionary class struggle into first a socialist, and ultimately a communist society. In this sense, then, he was looking forward to the new order rather than backward to the intersection of capitalism with previously existing formations (see especially Marx's last major work, *Das Capital*).

Second, as argued in the first two chapters here, Marx tended to equate the "free" wage labourer in the capitalist mode of production with the European working-class male. But this proletarian is himself the product of a specific Western European historical process, marked not only by the transition from feudalism to capitalism but also by a series of struggles with other groups of labourers. In this sense, then, he is not a "model" of the proletarian as such but is rather a privileged member of a working class in which he has sought to protect his wages and his position, often at the expense of other proletarians: women, children, non-European labourers, etc. Thus, his struggles involve not only or simply the capitalist class but also groups differentially situated within the working class, including groups of labourers who continue to live in societies or communities that intersect but are not identical with the capitalist economy. To miss this point is to miss much of the internal struggle that has characterized the UFAWU since its formation and that also influenced salmon canners. They hired specific groups of shoreworkers for specific tasks, establishing a division of labour that mirrored not only the industrialization of fish processing but also existing divisions according to the gender and race, as well as the age, of the group hired at any particular point in time.

Although I have argued that Marx's labour theory of value has serious limitations, I have not argued for its rejection, as have a number of Marxist and non-Marxist writers. Indeed, some consider this entire branch of Marx's analysis to be outdated and not worthy of study. For example, Giddens (1983b, 1–2, 20–1), in rejecting Marx's evolutionism, also rejects the notion of class as applying to any social formation other than capitalism. He also rejects the concept of mode of production because of the difficulty of applying Marx's own rather ambiguous schema (tribal society, ancient society, feudalism, socialism, and especially the problematic "asiatic" mode of production). Instead, Giddens proposes his theory of structuration as a way out of the problems he identifies with Marx's theory.

Unfortunately, in proposing the duality of structure, the interrelationship between agency and structure, Giddens does not overcome the dualism he identifies in existing social theories, including Marxist analysis. In his critique of class relations, Giddens (1983b, 81) reduces class to "the capital/wage-labour relation." While I would agree with Giddens in rejecting the evolutionism that mars Marx's analysis, I would argue that the vitality in his analysis of class lies not in specifying groups of people as they relate to each other in the economic sphere but in exploring the dialectic between use value and exchange value as these are played out in various social formations, whether capitalist or non-capitalist. And here I would argue not in favour of rejecting Marx's analysis but rather in favour of arguing that Marx was not sufficiently dialectical in his own insights on the interrelations between use value and exchange value. In the *Grundrisse* (Giddens only analyses one small section of it, the *Formen*), Marx gives primacy of place to exchange value in its transformative powers and the emergence of the capitalist economy. He ignores use value and sees the social formations in which it predominates as instances of "man's" subordination to the laws of nature which "he" is not yet in a position to comprehend and to master. I argued in chapter 2 that production for use also entails complex relations of equality and of domination and oppression and I agree with those feminist analysts who point out that production for use is always and everywhere accompanied by sharing or by exchange of some kind.

To use an example of the most basic kind, a mother shares her breast milk with her baby and this act is not only social but can also become a form of exploitation, as when wet nurses are hired to feed rich women's babies. The mother produces milk (and the production is not simply or only a biological act since women can and do control its production) which has use value and is exchanged with another human being (this defines its use value). The act of exchange is not limited to the mother/child dyad but has everywhere and always (as far as archaeological and anthropological evidence has been able to ascertain) defined the nature of the relationships among members of human groups. Even the most male-dominated theories which assume "man the hunter" as establishing human organization take for granted that the meat men bring from the hunt is used to feed nursing women and their children.

Hosts of social relations have sprung around this one act throughout history and in all societies, running the gamut from glorifying the power of women in their ability to nurture humans to vilifying and cursing them as dirty and defiled because of this same power. In "civilized" societies this act is taken for granted and rendered invisible, as are women themselves. In the capitalist mode of production, not only has breast milk been replaced and commodified by artificial substitutes but motherhood itself is becoming commodified through such practices as making contracts with surrogate mothers who are paid to bear babies for others.

In expanding the dialectical interplay between use and exchange values, class relations become one instance of inequality between groups of people. Other instances involve all the possibilities of categorizing humans into groups who are ranked as having or not having power in relation to one another. I have suggested that the concept of patriarchy must be studied in this way, as the process whereby "woman" became a category and all those who occupied it (occasionally men were included, including Chinese male shoreworkers defined as a "feminine race") were defined as not only powerless but also invisible, not fully human. Race was another category that emerged with colonialism and was used for the same purpose, to deny humanity to anyone who was not "white." The concept of ethnicity introduced a further distinction to differentiate, for example, the British from Europeans like Italians in the fishing industry. These categorizations have in turn been used within capitalism to create what I have called "cheap" wage labour. The exchange of labour for a wage is not only or simply determined by impersonal market forces. Whose labour is being bought figures in how that person or group is paid. The membership of the person in the designated group – immigrant, a woman, a child, Chinese, aboriginal, etc. – also influences the type of life that person leads. But this is not to say that change is not possible. Much of the content of the last few chapters has served to demonstrate how various disempowered groups in the fishing industry fought for and won recognition from capitalist employers, from union representatives, from fellow shoreworkers, and from members of their own families.

I agree with Giddens that a class analysis in its most narrow Marxist interpretation is insufficient for grasping the complexity of current class and group struggles. But the solution is not to erect yet another abstract schema and interpose the "duality" of structure, for such an analysis also misses the fact that it is not individuals as such who act, but individuals situated in a highly diverse and often conflicting set of groups. For example, a Mohawk may be a member of the Warrior society, part of the community of Kanesatake, a clan mother, a status Indian as defined in the Indian Act, a proletarian, etc. This person cannot be reduced to any one of these roles, but neither is she an individual who creates and reproduces the various structures in which she participates and/or which determine her future and her fate. Rather she acts within various groups and communities, and her ability to change her society depends on the nature of her membership and participation in these groups. It seems to me that Giddens misses these intervening group relations and leaves us with the actor on one side and the structure on the other. Of course, the more pertinent problem with his analysis is that he remains at the level of theoretical abstraction and never provides a historical analysis of any one social formation to demonstrate how his theory of structuration could actually be used in social analysis.

"Only if historical materialism is regarded as embodying the more abstract elements of a theory of human *Praxis*, snippets of which can be gleaned from the diversity of Marx's writings, does it remain an indispensable contribution to social theory today" (Giddens 1983b, 2). I have tried to push this conclusion in the opposite direction. What we need is not more grand theory but analysis that takes into account the diversity and complexity of lives as people live them from day to day. This is the feminist project – not to discover *the* one theory that will explain everything, but to bring forward diversity at all levels of analysis, using theoretical understanding as a way into the complexity of the social world in which we live.

There is another fundamental problem with Giddens' analysis which can also be found in other highly abstract theorizing. This is the tendency to use terms and concepts that have specific histories linked to the emergence of capitalism as if they are universal and historical, terms like "civilization," "ideology," "individual" (including "actor"), not to forget the term "history" itself. These terms have taken on a certain meaning that does not show them to be universal but rather particularly situated in Western modern thought. Even the terms "woman" and "man" have shifting meanings that relate them to particular periods of time and places. This is all the more peculiar for an author of the stature of Giddens given his own insistence on incorporating what he calls "time-space distanciation" into social theory and his earlier works in which he has showed far more sensitivity to the various philosophic traditions that have spawned many of these concepts.

Finally, many critics who reject the labour theory of value tend to invoke the base/superstructure dichotomy. One of the chief criticisms is that Marx was overly concerned with the economy as "determining" all other relations. While such an argument makes sense within capitalism where, for the first time in history, economic considerations are given primacy of place, it makes less sense in other types of social formation. But surely such an argument can be turned on its head. If economic relations in other social formations are embedded within other relations, then concepts like "use value" and "exchange value" must be divorced from a strictly economic determinism and understood rather as part of other relations, of other ways of "being" in the world.

Arendt criticized Marx for giving primacy to the necessity to labour, arguing instead that necessity itself was judged by the Athenian citizen of classical Greece as making "man" nothing more than an animal. It was his divorce from the necessity to labour (by forcing others to perform it for him in the privacy of his household) that qualified a man for *polis* life, for citizenship in a world erected by "man" against the world of cyclical nature. For Arendt this is the birth of a truly human political life that was destroyed in modern times by the reintroduction of necessity and labour as the underpinnings of capitalism. She notes that the genius of Marx was to identify this new development, but that his error was to then assume that labour and necessity have predominated throughout time. But the need to labour, for oneself and for others, remains a basic human condition. How that need is met, and by whom, is what changes. We cannot abstract backward from capitalism and find all traces of the present as inexorably prefigured in the past (this is to fall into the evolutionism that Giddens rightly points to in Marx's work). But this does situate the human capacity to labour as one capacity that is related to the others and that remains fundamentally important to human survival. We cannot reject it or simplify it but must search for the complex ways in which it has been used to define what is considered to be the human condition.

We would do well in this connection to study more carefully aboriginal systems of exchange/Exchange and of government. Certainly many of the peoples occupying what is today British Columbia also had violent practices involving warfare and subjugation of certain groups of people including, in some but not all cases, women. However, given the length of their history of occupation of their territories compared to those of the European colonizer, they were certainly far more successful in "managing" the fisheries resources then were the groups who took control after Confederation. What is needed is ongoing dialogue in which the voices of all those who are affected are heard, and negotiation that allows the empowerment of all groups as true equals in whatever decisions are made.

APPENDIX

Labour Classifications, Wage Scales, Conditions of Employment, Board and Accommodation Agreements for Canneries Operated by J.H. Todd & Sons Ltd, 1947–49

The years 1945 to 1949 marked an important transition for British Columbia shoreworkers. Federal government legislation enacted after the Second World War provided mechanisms for the legal recognition of industrial trade unions like the United Fishermen and Allied Workers Union. Legislation was also put in place forcing employers to list employees by name with salary, thus rendering obsolete the Chinese contract system as the major method of payment for the bulk of the seasonal labour forces. Nevertheless, the Chinese contract system continued as an important method of recruiting labour for several decades to come.

The UFAWU began the arduous task of certifying plants by signing on the permanent (white male) labour forces in the reduction plants and canneries. In addition it also began, with the Native Brotherhood of British Columbia (representing aboriginal fishers and shoreworkers), to organize the seasonal labour forces. It did so by using the categories of race and gender already in place to designate cheap wage labour. While the UFAWU (and the NBBC) were now the formal representatives of union fishers, tendermen, and shoreworkers, the companies were organized into the Salmon Canners' Operating Committee (SCOC), in existence for the better part of the century. The SCOC became the employers' negotiating arm with the UFAWU in establishing province-wide agreements. Initially, these agreements covered specific groups of labour and were organized into Supplements. The first agreements covered only white male labour classifications in the reduction plants and canneries. In the last years of the 1940s, the UFAWU was seeking to incorporate aboriginal and white women and Chinese male labourers by establishing separate supplements listing their classfications and methods of payment (monthly guarantees for men, overtime, hourly rates of pay and piece rates including overtime for women).

In its negotiations with the UFAWU, the SCOC prepared a series of questionnaires covering the three seasons from 1947 to 1949. The completed questionnaires for the three canneries owned by J.H. Todd & Sons Ltd (Inverness, the oldest cannery on Skeena River, Klemtu on the central coast, and Empire on Vancouver Island) have survived in company files currently stored in the Special Collections division of The University of British Columbia Library. Empire was also canning herring during these years. The information covered the cannery operations and included cannery labour as well as networkers.

I have incorporated as much of the information contained in those questionnaires as possible and combined them for the three canneries. What is striking is that the Male Cannery Workers' Supplement is highly detailed in its listing of job classifications and rates of pay for very few men (and that it is further subdivided into various groupings), while aboriginal women are lumped together with few job categories and in some cases (as in 1949) with no rates of pay indicated. It is also worthy of attention that each supplement covered a specific race and gender. The networkers supplement was broken down by "native," "white," "male," and "female," as was the information for board and accommodation. In 1947 there was a special supplement for Chinese men. By 1949 this supplement had become an appendage to the Male Cannery Workers Supplement as "Group 4 – with guarantee" and "Group 4 – no guarantee." With the abolishing of the Chinese contract, employers were attempting to convert the seasonal guarantees of their Chinese labour into monthly guarantees or even simply straight time hourly rates. The Chinese labour force resisted. For example, in 1949 at Klemtu the entire Chinese labour force refused to accept these conditions and the manager was forced to revert to the 1948 agreement. In the notes that accompanied the completed questionnaires, the managers expressed considerable hostility toward the Chinese, and one manager offered as a solution to smoothing relations with this sector the following cryptic comment: "Shoot the bastards." The result was the almost complete displacement from the industry of Chinese male labour.

The most crucial and skilled components had already become incorporated into machinery. Readers' attention is directed to the number of classifications with the word "Iron Chink" in them. These automatic butchering machines were maintained by white men covered by the first supplement and fed by the Chinese labourers they displaced. When I toured CANFISCO's Home Plant (located on the Vancouver wharf just outside China town) in the early 1980s, I was amazed to find Chinese men still feeding these mechanical monsters. When I asked the foreman why Chinese men were employed on this particular task (I found them concentrated in this one area), I was told the job was so boring and dangerous that when white men were assigned they tended to lose concentration, and resulting injuries often leading to loss of fingers. The foreman seemed to assume that there was something in the nature of these Chinese men that suited them for this particular set of tasks.

White men covered by the first supplement were guaranteed monthly rates of pay in 1947 and by 1949 had seen their work day shortened with overtime rates introduced as part of their contract. Meanwhile, women continued to be hired by the hour and/or worked on piece rates. It is evident that the bulk of Todd's female labour force was aboriginal and that child care was an important concern. In summary, then, this was the starting point of UFAWU shorework organizing, a starting point that represented a gender and racial hierarchy in place since the last quarter of the nineteenth century. Since archival information from J.H. Todd & Sons Ltd. is being used here, readers may be interested in knowing a little more about this company. Biographical sketches further illustrate the gender, ethnic, and racial categorizations that marked the origins of the salmon canning industry.

JACOB HUNTER TODD

In his notes (c. 1925), Henry Doyle describes Messrs Todd & Sons as the "oldest company in the province, having [acquired] its first cannery [the Richmond], on the Fraser River, in 1882. It is a private company, immensely wealthy, and its brands are well established and favourably known throughout the British Empire. Next to the BC Fishing and Packing Co. [BC Packers], it enjoys the largest trade in Canada" (Newell 1989, 220). In addition, William Todd and Robert Johnston (manager of the Todd canneries in northern British Columbia) in 1917 formed another company named Provincial Canning Co. Ltd, to operate the Provincial cannery on Rivers Inlet.

Todd was the only canner in British Columbia able to obtain permission to consistently operate fish traps on the Juan de Fuca Straits, supplying Empire Cannery located near Esquimalt on Vancouver Island. Todd's Scottish parsimony was legendary. McKervill (1967, 51–2) recounts:

> Another well known character among the canners was Jacob Todd. Jacob Hunter Todd came to British Columbia from Eastern Canada through the Cariboo country as an itinerant trader. When the Cariboo excitement was over he crossed to Victoria, entered the wholesale grocery business and made a fortune in a very short time. In the process he acquired the reputation of being shrewd even beyond the normal run of Scots merchants, a repuation which he brought with him into the canning business when in 1882 he built Richmond Cannery on the North Arm. This pawkiness lent point to more than one tale in which Todd figured.

McKervill goes on to recount the tale of how Jacob Todd "commandeered two Indians who fished for him to row him from Richmond Cannery up to New Westminster." It took some time and effort and the "lads" were looking forward to a good meal and hotel room when they arrived. Instead, Todd purchased a packet of hard-tack from a grocery store they were passing, "drew out four pieces and gave two each to the lads who had worked so hard to transport him up the river," and told them that was to be their supper. "Doyle further records that when the news reached the cannery, there was ugly grumbling and strike talk among the workers, and indeed there might have been serious trouble if the local manager, unknown to the owner, had not paid the men handsomely under the camouflage of 'fish prices.'" Apparently Jacob's son Charlie was "just as mean and tight" as his father (1967, 53).

Doyle (Newell 1989, 221) provided the following summary:

> While very close and anxious for profit themselves, they apparently hate to see anyone else make money. When the British Columbia Packers Association was formed, they verbally agreed to enter the amalgamation; but once its formation was assured, they repudiated their word and remained free lances. Recently, like their competitors, they made heavy losses and expressed themselves anxious to sell out. But when solicited to amalgamate with others, they declined to join, in the belief that with the others amalgamated, but as independent operators themselves, they could repeat their former successes.

Doyle is referring to the two periods when B.C. Packers attempted a major buy-out, in 1902 and again in 1928, to create a monopoly-like position (discussed in chapter 3). Todd & Sons weathered the storm and B.C. Packers had to wait another quarter century before it could absorb Todd's canneries. In the mid-1920s, Doyle noted that Charles F. Todd and his sons had continued the company begun by J.H. Todd. He describes Charles F. as "a man of about 70 years of age, but active and shrewd in business. He has few friends, and is simply a money-making machine – albeit a very successful one" (Newell 1989, 221).

In the sketch prepared for the *Dictionary of Canadian Biography*, Anderson notes that Jacob Hunter Todd, born in 1827 near Brampton, Upper Canada, was the son of John Todd and Isabella Hunter. "According to descendants of Jacob Hunter Todd, his father ... [a] farmer who spent more time fox hunting than farming, immigrated to the United States in 1816 and was joined by his wife two years later. He worked in New York City at several trades before moving in 1820 to a farm in Trafalgar Township, Upper Canada, preferring, in the words of the *Canada Christian Advocate*, 'life under British rule'" (1990, 1062–4).

Jacob Hunter Todd married Anne Fox in 1854, and he moved to Victoria in 1862, the year the town was incorporated, and Anne joined him two years later. In undertaking to build a fence around his property, he met Governor James Douglas. "Douglas was from the outset well disposed toward Todd, regarding him as a man of intiative, attached to British institutions." In 1864

Todd established a general store in Barkerville, "the heart of the gold-mining community." The town was destroyed by fire in 1868 and Todd suffered large losses (about $10,000 in goods and buildings) but rebuilt. "His store flourished, and he was willing to grub-stake miners and prospectors in whom he had confidence." When the Cariboo gold rush began to abate, he established, in 1873, new enterprises on Vancouver Island and on the mainland. "On Wharf Street in Victoria he opened a dry-goods and outfitting store that served as ship-chandlery to many sealing vessels, and he also operated a grocery store on lower Yates Street. In 1875 he set up J.H. Todd, a wholesale grocery firm, to supply goods to the interior of the mainland. Two years later he took his son Charles Fox into the company, renaming it J.H. Todd and Son" (Anderson 1990, 1062–3).

But Anderson notes that the most important of his ventures after he left the Cariboo was his entry into salmon canning. "J.H. Todd and Son's trademark 'Horseshoe' brand was the first to be registered (1881)." Eventually the company owned five canneries. The largest was purchased in 1906, Empire at Esquimalt. "Todd was also one of the first to establish large-scale commercial fish traps. Of the four ultimately operated by his company, the one near Sooke survived until the mid 1950s, probably the only early commercial trap on the coast to last so long." Jacob's other son Albert Edward also joined the firm. He sent his sons to school in Upper Canada. In his letters to them he admonished them to "behave in a pious and proper manner and, in particular, to refrain from reckless expenditure" (Anderson 1990, 1063).

"Markets were developed in Great Britain and Europe, and his 'Horseshoe' brand won prizes at London's Crystal Palace exhibition and at other world fairs. Through the efforts of the Todds as well as others in the industry, canned salmon became a popular food, particularly in Great Britain, where it was known as 'the working man's feast.'" Todd reinvested profits made from salmon canning into real estate. And he also became involved in politics at the federal and municipal levels. "After his death in 1899 it was reported in the press that his estate was the largest ever probated in British Columbia." And in his obituary the *Daily Colonist* noted that he was "one of those who made Victoria her reputation as the commercial center of the Pacific Northwest" (Anderson 1990, 1063–4).

Further confirmation of Todd's character emerges in the testimony he gave in 1892 before the British Columbia Fishery Commission. One of the commissioners questioned his statement that at one time offal had been "deposited in pens or cribs by order of the Department." Fisheries Inspector McNab was asked to verify this, which he did, although he pointed out that this had happened before his term in office. "Mr Todd. – When I make a statement, Mr Commissioners, you will please take it. (Applause from audience.)" When Wilmot had restored some semblance of order, he confessed to Todd that he had not known "that such was the case. – A. Yes, and it just shows the folly of having to defer to people at Ottawa who are so far away, they do not know anything about the matter" (BCFC 1892, 301). Todd also provided a detailed list of the catches of the boats contracted to fish for Richmond Cannery in 1888 and 1889, as well as Beaver Cannery for the years 1889 to 1891. The men fishing for him on these boats were paid for the fish brought to the plants and included "whitemen [*sic*], Indians, and all kinds, more or less." He saw the greatest competition coming from Alaska where 34 canneries had been built (302).

J.H. TODD & SONS LTD

The greatest amount of information available in the company files donated to Special Collections in The University of British Columbia Library covers the last years of the company's existence, the late 1940s and early 1950s. Having weathered so many storms in previous decades, it paradoxically seemed that it was the wartime boom years that precipitated the company's demise. In 1951 the company was celebrating its seventieth anniversary (1881–1951), a bittersweet celebration since three years later it ceased to exist.

As of 1953, the company still owned six canneries. Only three plants operated canning lines from 1947 to 1949: Klemtu, Inverness and Empire. Inverness was closed in 1950, after seventy-four years in operation (Blyth 1991, 119). Doyle had the following to recount about Klemtu: "The China Hat cannery (owned by Toms, Morris & Fraser) was only two years old when I bought it for British Columbia Packers Association. My associates, never regarding the purchase favourably, almost immediately dismantled and abandoned the plant. Subsequently, Morris became a cannery manager for J.H. Todd & Son and persuaded them to build a cannery near China Hat. It has been a heavy and consistent money-maker for this company" (Newell 1989, 60). Klemtu was closed in 1968, having operated for approximately forty years.

The termination of guaranteed packs secured by the British ministry of food appeared to have left canners with a huge inventory and difficulty re-establishing markets for their brands. In 1954 the British ministry of food agreed to purchase canned salmon, not to exceed $5 million from ABC, B.C. Packers, Canadian Fish, Cassiar, Great West Packing, Johnston Fishing and Packing, Francis Millerd, Nelson Brothers, Queen Charlotte, and J.H. Todd & Sons Limited (all members of the SCOC, now re-named the Fisheries Association of British Columbia). General Manager P.E. Paulson observed in a letter dated 1 March 1954:

> It has been a great help to our industry to be relieved of a certain amount of inventory through the purchase by the British Ministry of Food. Frankly I am certainly looking forward with great enthusiasm to the possibility of re-establishing the old time commer-

cial connections in the U.K. I always feel that these bulk purchases have a definite disadvantage in that your business is everybodys [*sic*] business and it makes it extremely difficult in dealing with labour, fishermen, etc. We all know that these bulk purchases are never big money making ventures, but it is difficult to put that over as an argument when discussing it with pressure groups such as I have mentioned. (Box 3, file 18)

It appears that Todd had been aggressively advertising in order to re-establish its own brands on Canadian grocery shelves when the order came through from the British food ministry. Todd might have already sent out packs to outlets, but poor sales left a huge inventory that it could not sell. Sales manager F.S. Dawson observed in a letter dated 14 September 1953 that "the Industry at the present time has produced canned salmon to a value of many millions of dollars and of course all Operators are urgently in need of cash sales business ... If we get too much salmon under our own labels at key points throughout Canada and sales volume does not materialize as expected, we can have seriously impaired our chances for a successful operation" (Box 5, file 30).

In 1954 the company closed its office in Victoria and consolidated its operations in Vancouver. There was a change in personnel, and the sales office was also closed. A demand loan was issued and new principals took control. The pack was to be sold on a bulk basis to Canadian Fish and BC Packers. Dawson wrote a customer in Montreal in a letter dated 28 June 1954: "I have, however, the unpleasant task of telling you that the operations of the Todd firm have been re-organized and new principals are now in control. The Todd company, with the advent of new pack supplies, will not market canned salmon under its own brands, nor, as I understand it, in its own name" (Box 3, file 41). The sale of Todd brand canned salmon ended on 30 June 1954 (Box 3, file 43; see also Box 5, file 33).

The absorption of Todd by the Big Two was kept secret for a number of years. The importance of brand names in an industry where the product tends to be fairly uniform became evident when Weston acquired BC Packers, a major reason being the acquisition of its Clover Leaf brand of canned fish which it could now carry exclusively in its Loblaw grocery chain stores across the country (Davies 1987). But by the end of the 1940s, the dynamism of the Todd family empire seems to have been on the wane. Profits from the fisheries continued to be invested in real estate, and the canneries were increasingly being operated by managers who were not family members. The Todd family did not appear to be very anxious to continue in this line of business, given its ups and downs and the need to continually invest much time, energy, and money in building, machinery, operations, labour, and marketing.

It would also appear that the heyday of the one-man or one-family operation had passed. Millerd (1988, 36) notes that company mergers and technological change after 1928 led to concentration of processing operations in urban centres: "No longer was there a need to have a cannery adjacent to every fishing ground." Thus the advantage the Todds had acquired in possessing canneries on major salmon producing streams was being eroded as fishing vessels and tender fleets were developed capable of transporting raw fish over longer distances. By 1945 only 29 canneries were in operation (table 7). And between 1943 and 1955 the shoreworker labour force was almost halved, from 6066 to 3587 labourers (Warriner 1987, 330, table 14.1).

With further corporate concentration (and Todd was a casualty) and amalgamation of processing operations to Prince Rupert and the Greater Vancouver Region, coastal communities for whom cannery work was crucial saw their canneries closed. Paradoxically, as air, road, and tug and barge services improved and canneries were shut down, coastal communities were left with very poor transportation services (Millerd 1988, 37). Aboriginal peoples suffered most acutely.

> The up-coast Indian is only three generations removed from the day when the whole coast was his to fish. In three generations the Indians have learned, with varying degrees of success, to accommodate themselves to the white man's world of money and machines, liquor and licensing, property and regulations. They have been subjected to the exploitation of the traders, the experiments of the government and the strictures of ultra-evangelistic Christianity. They fished for the canneries in the infancy of the industry before white men and Japanese came with their nets to share in the silvery wealth of the sea. Now their fishermen are outnumbered. Favourite drifts taught to them by their fathers are monopolized by the rapacious fleet from the big city and the jocund days when Louie Hall and his crew from the village of Klemtu could scoop up 47,000 humpbacks in one set are gone forever. Many inlets and channels that once teemed with fish, providing whole villages with food year after year, are closed indefinitely to fishing due to the near depletion of the stocks.
>
> Once the Indian women worked in the draughty, shed-like canneries that cluttered the beaches and the bays. Now with centralization of the industry in a few major spots and perfected automation that can spot out 250 cans per minute from each of as many as five lines operating simultaneously, few Indian women are needed. (McKervill 1967, 151–2)

LABOUR RELATIONS 1948 QUESTIONNAIRES

(compiled by the Salmon Canners' Operating Committee for 1947 salmon season)

These tables were compiled from completed questionnaires sent by the various plant managers for the Todd canneries, in response to letters from the Salmon Canners' Operating Committee requesting data in preparation for negotiations with the UFAWU in the late 1940s. For example, the covering letter dated 21 January 1948 listed the following companies in its letterhead: A.B.C. Packing Co. Ltd, B.C. Packers Ltd, Canadian Fishing Co. Ltd, Cassiar Packing Co. Ltd, Colonial Packers Ltd, Fraser Valley Canners Ltd, Great West Packing Co. Ltd, Johnston Fishing & Packing Co., Francis Millerd & Co. Ltd, National Fisheries Ltd, Nelson Brothers Fisheries Ltd, North Shore Packing Co. Ltd, Queen Charlotte Fisheries Ltd, J.H. Todd & Sons Ltd. All these companies were negotiating with the union through the Salmon Canners' Operating Committee. J.H. Todd & Sons sent each operating cannery manager a questionnaire to be completed and sent back to the head office in Victoria. The tables I have compiled here are based on the questionnaires completed for individual canneries as retained in the archival records. Since the classifications and rates of pay (unless otherwise indicated) remained the same, I combined the data across the canneries: Klemtu, Inverness, and Empire. Blanks indicate that no information was recorded and unless otherwise specified "–" indicate the marks as placed on the forms by the person completing them. All this data can be found in Box 21 of the Todd collection housed in Special Collections in The University of British Columbia Library.

1. Male Cannery Workers' Supplement

GROUP 1: MACHINE OPERATORS

Classification	*Rate of Pay (monthly)*	*# of Employees at Following:*		
		Klemtu	*Inverness*	*Empire*
Lineman	$225	1	1	1
Reform Man	$215	1	1	
Iron Chink Man	$215	1	1	1
Filling Machine Man	$215	1	1	1
Heading Machine Man	$175			
Port Engineer &/or Machinist	$225	1	1	
Ass't Port Eng./Mach.	$200		1	
Labelling Machine Man	$175			
Junior Machine Helper	$125			
1st Year Ass't Machine Man	$154		1	
2nd Year Ass't Machine Man	$170	1		
3rd Year Ass't Machine Man	$185			
Qualified Ass't Lineman	$200	1		
Qual. Ass't Reform Man	$190			
Qual. Ass't Iron Chink Man	$190			
Qual. Ass't Filling Mach. Man	$190	1		
SUB-TOTALS		8	7	3

GROUP 2: BOILER HOUSE

Classification	*Rate of Pay (monthly)*	*# of Employees at Following:* Klemtu	Inverness	Empire
Chief Eng. 2nd Class Ticket	$225			1
Chief Eng. 3rd Class "	$185			
Engineer & Retort (Dual)	$225	1	1	
Engineer & Fireman	$225			
Shift Engineer	$190			
Fireman: Hand Fired Boilers	$175			
Fireman: Coal Stokers & Oil Burners	$165	1	1	1
SUB-TOTALS		2	2	2

GROUP 3: FISH WHARF

Classification	*Rate of Pay (monthly)*	*# of Employees at Following:* Klemtu	Inverness	Empire
1st Year Talleyman	$154			
2nd Year Talleyman	$170			
Qualified Talleyman	$180	1	1	1
Winchman	$170			
Marine Leg Man	$170			
SUB-TOTALS		1	1	1

GROUP 4: SUNDRY

Classification	*Rate of Pay (monthly)*	*# of Employees at Following:* Klemtu	Inverness	Empire
Retort Man	$180	1		1
Oil Plant Operator	$170	1		
Ass't Oil Plant Operator	$154			
Maintenance Men:				
Carpenters, Blacksmiths, Electricians	$200	1 carp. 1 elect.	1 carp.	
Jeep Drivers (experienced)	$170	1	1	
Casual Labour (monthly)	$150	1		
Casual Labour (per hour)	75 cts	5		3
Watchman (7% increase over 1947)		1	1	1
SUB-TOTALS		12	3	5

SALMON OPERATION:
ADDITIONAL CLASSIFICATIONS

Classification	*Rate of Pay (monthly)*	*# of Employees at Following:* Klemtu	Inverness	Empire
Ice Engineer (cents/hour)	97 cts	1		
Camp Net Man	$210	1		
Net Boss	$295	1		
Cold Storage Engineer	$225	1		
Caretaker	$150	1	1	1
Storekeeper	$170			
Net Man	$220			
SUB-TOTALS		5	1	1

2. *Women Cannery Workers' Supplement*

Classification	*Klemtu*		*Inverness*		*Empire*	
	white	*native*	*white*	*native*	*white*	*native*
Slimers				19		10
Washers	7	20				8
Line Workers		8		6		5
Filling Machine Feeders		6		8		6
Index Feeders						
Warehouse & Can Loft	2	3		10		6
Labelling Machine Operator						
Lidding Mach. Operator						3
Packing Room Machine Lineworkers		4				5
Salmon Line Punchers						
Butchers						
Beginners						
Fillers						5
SUB-TOTALS	9	41	0	43	0	48

*3. Networkers Supplement**

Classification	*Rate Of Pay (1948)*		*# of Employees at (1947):*			
			Klemtu		*Inverness*	
a) monthly rated:		*overtime*	*white*	*native*	*white*	*native*
Seasonal Net Boss	$315	nil	1		1	
Yearly Net Boss	$235–$280	$1.22–$1.45				
Assistant Net Boss	$240	$1.25	1			
Combination Camp & Net Man	$235–$245	nil	1		1	
Camp Net Man	$220	nil				
Qualified Net Man	$225	$1.17	1		2	
SUB-TOTALS			4	0	4	0
b) hourly rated:						
Fisherman Qualified Net Man (temp. empl'd net work)	$1.17	$1.75.5		1	5	
Semi-Qual. Net Man	.95	$1.42.5				
Inexper. Net Man	.80	$1.20			1	
Qualified Net Man**	$1.17	$1.75.5		2		
Semi-Qual Net Woman	.80	$1.20		1		
Inexper. Net Woman	.75	$1.12.5		2		
Qual. Net Woman	.95	$1.42.5				5
SUB-TOTALS			0	6	6	5

* Networkers' Supplement for Empire was filled in as "not applicable" (indicating no net operations there that season). The rates of pay for the 1947 season were not given but rates for 1948 are available and were used here to allow comparison (for example, note the discrepancy between the rates paid to qualified net men and qualified net women).

**The rate was not filled in for this classification but it is assumed that it would be equivalent to the monthly rates and to the classification "Fisherman Qualified Net Man."

4. *Chinese Cannery Workers' Questionnaire (1947 season)*

Classification	*Rates of Pay**				*# of Employees At:*		
	monthly	*OT*	*hourly*	*OT*	*Klemtu*	*Inverness*	*Empire*
Butchers	$163.20	.85	.85	$1.27.5	3	2	3
Iron Chink Feeders	$163.20	.85	.85	$1.27.5	2	1	1
Unloaders	$163.20	.85	.85	$1.27.5	7	5	4 ($150)
Box Makers**	$163.20	.85	.85	$1.27.5	2	3	
Gang Knife Feeders	$153.60	.80	.80	$1.20	2	2	1
Binmen	$153.60	.80	.80	$1.20	3	1	2
Cooler Men on Hot Trays	$153.60	.80	.80	$1.20	10	9	3 ($150)
General Labour	$150.00	.78	.75	$1.12.5	13	4	6
Pre-Season Work	$150.00	.78	.75	$1.12.5	12	15	
SUB-TOTALS					54	42	20

ADDITIONAL CLASSIFICATIONS

Warehouse Foreman	$194	none payable	1	1	
Cook	$180	none payable	1	1	
Assistant Cook	$155	none payable	1	1	
Time Keeper	$205	none payable	1	1	

* Chinese employees at Klemtu, Inverness and Empire were paid on a monthly basis. The general agreement among canning companies was that out-of-town plants paid their Chinese employees on a monthly basis while in-town plants could decide on an individual basis. The manager for Empire noted two discrepancies as shown. Board charges of $30/month were deducted for Chinese employees at canneries operating messhouses (all out-of-town plants and some in town). Overtime was defined as in excess of 8 hours per day, after noon on Saturday and all day Sunday except where the flexible weekend was in force: 36 hours consecutive rest. All pre-season work was paid the general labour rate until the canning season opened at which time labourers were paid according to the classification they occupied.

**Some Box Makers were also paid on a piece work basis at Klemtu at $2 per case and during the herring season at Empire.

LABOUR RELATIONS
1948 QUESTIONNAIRES
EMPIRE CANNERY: 1947 HERRING SEASON

1. Male Cannery Workers' Supplement
(only classifications used are listed here)

Classification	*Rate of Pay (monthly)*	*# of Employees*
GROUP 1: MACHINE OPERATORS		
Lineman	$225	1
Heading Machine Man	$175	1
1st Year Ass't Machine Man	$154	1
GROUP 2: BOILER HOUSE		
Chief Eng. 2nd class ticket	$225	1
Shift Engineer	$190	1
GROUP 3: FISH WHARF		
Qualified Talleyman	$180	1
GROUP 4: SUNDRY		
Retort Man	$180	1
Oil Plant Operator	$170	1
Jeep Drivers (Experienced)	$170	2
Casual Labour	.75/hour	78
Watchman	7% increase over 1947	2

2. Women Cannery Workers' Supplement

Classification	*White*	*Native*
Lineworkers		
Patchers		
Packing Room	6	6
Can Loft	12	4
Beginners		
Fillers	5	50
Machine Headers	13	7
Hand Headers		
Herring & Pilchard		
Line Punchers		
Hot Patchers	8	8
TOTAL	44	75

3. Chinese Cannery Workers' Questionnaire

Classification	*# Employed (All Monthly Rated)*
Box Makers	3
Binmen	4
Cooler Men on Hot Trays	6
General Labour	16
ADDITIONAL CLASSIFICATIONS	
Cook Messhouse	1 ($180)
Cook Chinese Mess	1 ($180)
Time Keeper	1 ($205)
Foreman	1
Packing Room Foreman	1 ($235)
Punchers	6 ($153.60)
Assistant Cook	1 ($150)

LABOUR RELATIONS 1948 QUESTIONNAIRES QUESTIONNAIRE RE: BOARD, HOUSING & SERVICES

A. Male (white):	*Klemtu*	*Inverness*	*Empire*
# lodged in bunkhouse & ate in messhouse	21	20	11
amount paid for meals per month	$30	$30	$30
messhouse cost per man per month	$64.35	$53.01	$49.50
average # of meals served peak season	90/day	60/day	35/day
approx. length salmon season	8 mos	7 mos	herring
COOK HOUSE (4 classifications listed)			
cook (Chinese males & white females)	1 Ch M	2 W F	1 Ch M
assistant cook	1 Ch M		
flunky			1 Ch M
wages for cook	$185	$150	$188
wages for assistant cook	$150		
wages for flunky			$120
HOUSES			
How many houses? (Klemtu: for both M&F)	6	5	no ans
Was rent charged?	no	no	
electric light (no charge)?	yes	yes	
fuel (no charge)? Klemtu: stove Klemtu + Empire: oil/coal; Inverness: coal/wood	yes	yes	

B. Female (white)

Inverness: ANSWER: "all natives, does not apply"
Empire: no answers given
Klemtu: Eight white women were lodged in the women's bunkhouse and ate in the messhouse. The women pooled and paid $27/month for board. The cook was a white female and was paid $120/month. An average of 24 meals per day served at peak of season and the approximate length of the salmon season was 3 months.

C. Chinese	*Klemtu*	*Inverness*	*Empire*
($30 per month deducted for board from wages)			
# lodged in bunkhouse & ate in mess	43	33	35
messhouse cost per man per month	$50.40	$56	no ans
average # of meals per day at peak	129	99	no ans
approx. length of salmon season	6 mos	2.5 mos	no ans
COOK HOUSE:			
cook	1 Ch M	1 Ch M	
cook's wages	$180	$180	
assistant cook	1 Ch M	1 Ch M	
assistant cook's wages	$155	$155	

D. Natives	*Klemtu*	*Inverness*	*Empire*
How many houses did you make available?	15	53	38
Did you charge rent?	no	no	no
Did you supply electric light?	yes	no	yes
type of fuel (no charge)	stove oil/coal	wood	not specified
	(limit of $12 per house)		

LABOUR RELATIONS QUESTIONNAIRES (1948 SALMON & HERRING)

Cannery Machinemen & Misc. Male Classification

Classification	*Rate of Pay (monthly)*	*Overtime (per hr)*	*# of Employees at:* Klemtu salmon	Inverness salmon	Empire salmon	Empire herring
GROUP 1						
Lineman	$245	$1.28	1	1	1	1
Reform Man	$235	$1.22	1	1		
Iron Chink Man	$235	$1.22	1	1	1	
Filling Machine Man	$234	$1.22	1	1		
Heading Machine Man	$195	$1.02				1
Port Eng &/or Machinist	$245	$1.28	1	1		
Ass't Port Eng/Mach	$218	$1.14				
Labelling Machine Man	$195	$1.02				
Junior Machine Helper	$140	$ .73				
1st Yr Utility Man	$170	$ .88				
2nd Yr Utility Man	$185	$ .96			1	1
3rd Yr Utility Man	$200	$1.04				
Qual Ass't Lineman	$220	$1.15	1			
Qual Ass't Reform, Iron Chink or Filling Machine Man	$210	$1.09				
Retort Man	$200	$1.04	1		1	1
Oil Plant Operator	$190	$ .99				
Ass't Oil Plant Oper	$175	$ .91				
Maintenance Men	$220	$1.15		3	1	1
GROUP 2						
Chief Engineer (2nd cl)	$245	$1.28				
Chief Eng (3rd Class)	$235	$1.22	1			
Eng & Retort (dual)	$245	$1.28		1		
Eng & Fireman (dual)	$245	$1.28			1	1
Shift Engineer	$210	$1.09			1	1
Fireman (hand fired boilers)	$195	$1.02				
Fireman (coal stokers & oil burners)	$185	$ .96	1	1		

GROUP 3	*monthly*	*O/T*	*K(sal)*	*I(sal)*	*E(sal)*	*E(her)*
1st Year Talleyman	$170	$ .88				
2nd Yr Talleyman	$185	$ .90				
Qualified Talleyman	$200	$1.04	1	1	1	1
Winchman	$190	$ .99				
Marine Leg Man	$190	$ .99				
Jeep Drivers (Exper)	$190	$ .99	1	1	1	1

GROUP 4 This group was broken down further into "white" (W) and "Chinese" (C).

Classification	*monthly*	*O/T*	*K(sal)*		*I(sal)*		*E(sal)*		*E(her)*	
			W	*C*	*W*	*C*	*W*	*C*	*W*	*C*
Butchers (hourly: $ .95; $1.42.5 O/T)	$183	$ .95				1				
Iron Chink Head Knife Feeders ($ .95; $1.42.5)	$183	$ .95		5		2		2		
Unloaders (salmon) ($ .95; $1.42.5)	$183	$ .95		7		5			3	
Boxmakers (also piece rates)	$183	$ .95		2		2				
Gang Knife/Iron Chink Feeders ($ .90/hr; $1.35)	$173	$ .90		3		1				
Binmen & Cutters ($ .90;$1.35)	$173	$ .90		6		6				12
Cooler Men on Hot Trays ($ .90; $1.35)	$173	$ .90		9		6		2		16
Casual Labour ($ .88; $1.32)	$170	$ .88		7	2	2		6	4	5
Empire: add 35 native men herring					for 10 hrs			hrs	hrs	

LABOUR RELATIONS QUESTIONNAIRES (1949) WOMEN CANNERY WORKERS SUPPLEMENT 1948 SEASON

Each cannery is broken down by race: White (W) and Native (N). Klemtu and Inverness had salmon operations. Empire had salmon and herring.

Classification	*Rate of Pay*		*# of Employees at:*							
			Klemtu		*Inverness*		*Empire*			
							sal		*herring*	
	hourly	*O/T*	*W*	*N*	*W*	*N*	*W*	*N*	*W*	*N*
Slimers, Washers, Lineworkers	$ .75	$1.12.5	4	23		30	3	6		
Filling Mach Feeders	.78	1.17		8		4	3	2		
Index Feeders	.75	1.12.5				2	4		9	9
Warehouse & Can Loft	.68	1.02	4	9		6				
Labelling Mach. Operator	.80	1.20								
Box Lidding Mach. Oper.	.80	1.20								
Herring & Pilchard Line Punchers	.80	1.20								
Packing Room Mach. Lineworkers	.73	1.09.5								
Salmon Line Punchers	.75	1.12.5						3		
Hot Patchers (herring/pilchard)	.80	1.20		4					6	8
Butchers	.90	1.30								
Beginners (1st 48 hrs)	.65	1.02.5								
Registered Nurses										

ADDITIONAL CLASSIFICATIONS: WOMEN CANNERY WORKERS SUPPLEMENT

Nurses were paid $ .75/hr (no O/T rate listed). For the classification Registered Nurses there was no rate given in the table. Klemtu had 4 native nurses and Inverness had 2 native nurses.

There were two piece work classifications: headers and hand fillers. Empire had 8 white women and 8 native women headers for its herring operation. Empire also employed 47 native women hand filling herring.

Piece Work Rates for Hand Filling Salmon: $ .21.7 per case of 96 half-pound flats or the equivalent of $11.85 per ticket of 150 trays of 35 half-pound flats or $ .15.8 per case of 48 one-pound talls or the equivalent of $11.85 for a ticket of 150 trays of 24 one-pound talls or $ .19.9 per case of 96 one-quarter pound flats or the equivalent of $12.45 per ticket of 150 trays of 40 one-quarter pound flats.

Piece Work Rates for Hand Filling Herring & Pilchard: $ .12.4 per case of 48 one-pound ovals; $ .14.6 per case of 48 one-pound talls; or 20 cents per case of 96 half-pound ovals.

Piece Work Rates for Heading Herring and Pilchard:
7.8 cents per box of 70 pounds by machine – heads only
8.6 cents per box of 70 pounds by machine – heads & tails
15.6 cents per box of 70 pounds – hand heading

Overtime for Piece Work: $ .37.5 cents per hour in addition

LABOUR RELATIONS QUESTIONNAIRES 1949

SOURCE: section entitled "Statistical Questions" asking for following information:
(1) # of employees employed each month & covered by agreement
(2) # of employees eating in the messhouse each month
(3) # of employees who paid UFAWU union dues by check-off each month
(4) # of native employees who paid Brotherhood or Sisterhood dues by check-off each month
If (3) or (4) not indicated in table because NIL response.

1. *Klemtu*

	Jan	*Feb*	*Mar*	*Apr*	*May*	*June*	*July*	*Aug*	*Sept*	*Oct*	*Nov*	*Dec*	TOTAL
NATIVE			51	19	12	20	4		10	22	1		139
Cannery Mach.													
(1)	9	9	11	11	11	11	14	12	9	14	2	2	115
(2)	4	4	7	8	7	6	11	9	5	5	–	–	67
(3)	8	8	8	8	11	10	11	10	9	9	2	2	96
Chinese:													
(1)						45	45	42	41	41	–	–	214
(2)						45	45	42	41	41	–	–	214
Women:													
(1)			45	–	–	–	51	63	59	15	–	–	234
(2)								1	1	1			3

Note: "–" here refers to no runs of fish.

281 J.H. Todd & Sons Ltd

Networkers: 2 employed Jan-Mar; 4 Apr-May; 3 in June; 2 July-Sept and 1 in Oct.
Native Networkers: 2 in May; 3 in June; 6 in July and 2 Sept/Oct.
One male paid UFAWU check-off dues Jan-Oct and another added for April to Sept incl. No females listed.
Fresh Fish & Cold Storage: 1 employed Jan to Sept and an extra male in Feb.
One paid UFAWU check-off Jan to June incl.
Store Clerks employed: categories 2–4 not filled in.
Male: 1 Jan-Mar; 2 Apr-May; 3 June/July; 4 Aug.; 3 Sept; 2 Oct; 1 Nov-Dec.
Female: 1 Mar-Apr; 2 May-Aug; 1 Sept.
Misc. classifications not covered by union agreements: oil station attendents, fire chiefs, bunkhouse caretakers, etc.
employed: 1 all year and a second Apr-July and one paid UFAWU check-off Jan to Apr and Nov-Dec.

2. *Inverness*

AGREEMENT		*Mar*	*Apr*	*May*	*June*	*July*	*Aug*	*Sept*	*Oct*	*Nov*	TOTAL
Cannery Mach.											
Men:	(1)		18	19	38	30	28	21	5		159
	(2)		8	8	12	12	10	9	2		61
	(3)		–	3	4	8	9	7	–		31
Chinese:	(1)				25	25	25	23	–		98
	(2)				25	25	25	23	–		98
	(3)				–	21	18	18	–		57
Women:	(1)				36	34	30	26	–		126
	(3)					2	2	2	–		6
Networkers:	(1)		7	15	9	12	11	9	–		63
(male)	(2)		4	5	4	3	5	3	–		24
(male)	(3)			1	1	1	2	2	–		7
store clerks:											
male	(1)	1	1	1	1	1	1	1	1		8
female	(1)				1	1	1	1	1		5
male	(2)		1	1	1	1	1	1	1		7
messhouse staff											
(female)	(1)		2	2	2	2	2	2	2		14

Misc. not covered by agreement: 2 employed Jan-May, 4 in June, 1 July-Oct and 2 Nov/Dec.

3. Empire

AGREEMENT		May	June	July	Aug	Sept	Oct	Nov	TOTAL
Cannery & Misc.:									
Native				5	5	5	35	35	85
White	(1)	4	4	11	13	13	17	17	79
	(2)						12	12	24
	(3)			1	3	3	8	8	23
	(4)						1	5	6
Chinese:	(1)				10	10	33	33	86
	(2)			10	10	10	33	33	96
	(3)						2	2	4
Women (White & Native):	(1)			23	22	22	84	94	245
	(3)				10	10	12	12	44
	(4)						4	35	39
Store Clerks:									
(male)	(1)						1	1	2
	(2)						1	1	2
Messhouse Staff									
(male)	(1)			1	1	1	3	3	9

QUESTIONNAIRES: 1949 SEASON CANNERY MACHINEMEN & MISC. MALE CLASSIFICATIONS

Classification	*Hourly Rate*	*Min. Monthly Guarantee*	*# Of Employees At: Klemtu*	*Inverness*	*Empire*
GROUP 1:					
Lineman	$1.33	$255	1	1	1
Reform Man	$1.27	$245	1	1	
Iron Chink Man	$1.27	$245	1	1	1
Filling Mach. Man	$1.27	$245	1	1	1
Heading Mach. Man	$1.07	$205			
Port Eng/Mach.	$1.33	$255	1	1	
Ass't Port Mach.	$1.19	$228		1	
Labelling &/or Box-up Machineman	$1.09	$210			
Junior Machine Helper	$ .78	$150			
1st Yr Utility Man	$ .93	$180	1		
2nd Yr Util Man	$1.01	$195	2		
3rd Yr Util Man	$1.09	$210			
Qual Ass't Lineman	$1.20	$230	1		
Qual Ass't Reform, Iron Ch. or Filling Mach.	$1.14	$220			
Retort Man	$1.09	$210		1	
Oil Plant Oper	$1.04	$200			1
Mtce Men	$1.20	$230		1	
GROUP 2:					
Chief Eng (2nd class ticket)	$1.33	$255	1		
Chief Eng (3rd)	$1.27	$245			1
Eng & Retort (dual)	$1.33	$255		1	
Eng & Fireman (dual)	$1.33	$255			
Shift Engineer	$1.14	$220			1
Fireman – hand fired boilers	$1.07	$205			
Fireman – coal stokers & oil burners	$1.01	$195	1	1	

GROUP 3:

Classification	*Hourly*	*Monthly*	*Klemtu*	*Inverness*	*Empire*
1st Yr Talleyman	$ .93	$180			
2nd Yr Talleyman	$1.01	$195			
Qualified Talleyman	$1.09	$210	1		
Winchman &/or Weighing Stn Scale	$1.04	$200			
Marine Leg Man	$1.04	$200			
Jeep Driver (Exper)	$1.04	$200	1	1	1

GROUP 4 – WITH GUARANTEE:

Butchers	$1.00	$193	4		
Iron Chink Head Knife Feeders	$1.00	$193	6		2
Unloaders (salmon)	$1.00	$193	7	6	2
Boxmakers (wooden or combined)	$1.00	$193	2		
Gang Knife & Iron Chink Feeders (Indexers)	$ .95	$183		7	2
Binmen & Cutters	$ .95	$183		7	
Cooler Men on Hot Trays (herring)	$ .95	$183	8		2
Casual Labour	$ .93	$180	7	5	2

GROUP 4 – NO GUARANTEE: Hourly rates: $1.02 for 1st 3 classifications; no listing for Boxmakers; 97 cents for next 3 classifications and 95 cents for casual labour. No employees listed here for Todd.

PIECE WORK: Carton Stitching: 75 cents per 100 (73 cents for helper) Empire and Klemtu: 88 cents/hour Boxmakers: $2.25 per 100. No employees listed.

SEPARATE LISTING FOR KLEMTU CANNERY

NOTE: "Chinese were paid on 1948 Agreement at Klemtu."

Classification	*Min Monthly Guarantee*	*# Employed*
Unloaders	$183	11
Iron Chink	$183	4
Retorts	$173	2
Casual Labour	$170	8
Fish Bins	$173	2
Gang Knife	$173	4
Clincher	$170	2
Coolers	$173	8
Lye Washers	$173	5
Mailers	$183	2
Packing Room	$173	2

WOMEN CANNERY WORKERS SUPPLEMENT (1949)

	# of Employees at:					
	Klemtu		*Inverness*		*Empire*	
Classification	*white*	*native*	*white*	*native*	*white*	*native*
Slimers, Washers		11		12	2	8
Lineworkers				6	2	3
Filling Machine Feeders			1	8	3	4
Index Feeders				2	1	3
Warehouse	3	5		5	5	
Can Loft	1	2	4		5	

ADDITIONAL CLASSIFICATIONS:
Native Nurse Maids (80 cents per hour): 2 at Klemtu and 3 at Inverness (1 white)
Patchers (80 cents per hour): 6 native Klemtu; 1 white & 2 native Inverness
Feeders (83 cents per hour): 9 native at Klemtu

PIECE RATE WORKERS:
1949/50 herring season at Empire:
96 cents per hour hand-filling ovals
$1.17 per hour heading machine
TUNA: 80 cents per hour – fillers (32)
NOTE: no rates indicated for other classifications

NETWORKERS SUPPLEMENT:
1 seasonal net boss monthly rated at Klemtu & Inverness (total: 2 white)
1 qualified net man monthly rated at Klemtu & Inverness (total: 2 white)
1 camp net man (white) monthly rated at Inverness

HOURLY RATED:
Qualified net man (temporarily employed on net work): 3 native at Klemtu and 1 white Inverness
Qualified Net Woman: 4 native at Klemtu and 10 native at Inverness

Bibliography

MANUSCRIPT SOURCES

B.C. Packers. 1937–49. *Annual Reports*. British Columbia.

Hotel, Restaurant and Culinary Bartenders Union, Local 28. 1933–48. Minutes of meetings. Vancouver: Special Collections, University of British Columbia.

Insurers' Advisory Organization. 1923. Folder 9-7: British Columbia Canneries, surveyed by British Columbia Underwriters' Association. Vancouver: Special Collections, University of British Columbia.

J.H. Todd and Sons, Ltd. 1943–1954. Company files. Special Collections. Vancouver: University of British Columbia Library.

United Fishermen and Allied Workers Union. 1945–58. Collected papers. Special Collections. Vancouver: University of British Columbia Library.

Vancouver Trades and Labor Council, 1933–48. Minutes of regular meetings. Special Collections. Vancouver: University of British Columbia Library.

GOVERNMENT PUBLICATIONS

CANADA

British Columbia. 1875. "Report of the Government of British Columbia on the Subject of Indian Reserves." In *Sessional Papers of British Columbia* 1876, 57–72. Victoria: Government Printing Office.

British Columbia. 1884. "Metlakatlah Inquiry. Report of the Commissioners together with the Evidence." In *Sessional Papers of British Columbia* 1885, 131–6 and i–lxxxii. Victoria: Government Printing Office.

British Columbia. 1888. "PAPERS Relating to the Commission appointed to enquire into the state and condition of the Indians of the North-West Coast of British Columbia." In *Sessional Papers of British Columbia* 1888, 415–62L. Victoria: Government Printing Office.

Canada. 1868–1925. Annual Reports, Supplements and Appendices of the Fisheries Branch of the Department of Marine and Fisheries. In *Sessional Papers.*

Canada. 1869. "RETURN to an Address of the House of Commons, For information relating to the issue of Licenses to foreign fishing vessels, and correspondence affecting the protection of the Fisheries of Canada." *Sessional Papers*, no. 12, 1869, 1–45.

Canada. 1870. "RETURN to an Address of the House of Commons, dated 3rd March, 1870, for Copies of all Correspondence with the Imperial Government, relative to the admission or exclusion of American Fishing Vessels from the waters of the Dominion, and all Orders in Council on the subject." *Sessional Papers*, no. 81, 1870, 1–19.

Canada. 1871. "Message. The Governor General transmits for the information of the House of Commons the accompanying papers relative to the proposed union of British Columbia with the Dominion of Canada." *Sessional Papers*, no. 18, 1871, 1–29.

Canada. 1872–1902. Annual Reports of the Department of Indian Affairs. *Sessional Papers.*

Canada. 1872. "British Columbia. Report of the Hon. H.L. Langevin, C.B., Minister of Public Works." *Sessional Papers*, no. 10, 1872, 1–246.

Canada. 1872. "Message, Despatches, and Minutes of the Privy Council, relating to the Treaty of Washington." *Sessional Papers*, no. 18, 1872, 1–67.

Canada. 1876. "RETURN for all correspondence respecting the non-admission of Fish Oils and Fish of all kinds, the produce of the Province of British Columbia, into the United States, free of duty, under the Treaty of Washington of May 8, 1871." *Sessional Papers*, no. 42, 1876, 1–11.

Canada. 1881. "Report of the Census of 1881." Ottawa.

Canada. 1883. "RETURN for Copies of all Correspondence of recent date between the Government of the Dominion, and the Government of British Columbia, and all Orders in Council on the subject of Immigration into that Province." Hector L. Langevin (Acting Secretary of State). *Sessional Papers*, no. 93, 1883, 1–19.

Canada. 1885. "Report of the Royal Commission on Chinese Immigration. Report and Evidence." *Sessional Papers*, no. 54a, 1885, i–cii (Report); 1–487 (Minutes of Evidence).

Canada. 1893. "Census of Canada 1890–91." vol. 1. Ottawa: S.E. Dawson, Printer to the Queen's Most Excellent Majesty.

Canada. 1893. "British Columbia Fishery Commission Report 1892." *Sessional Papers*, no. 10c, 1893, i–xiii (Report); 1–433 (Record of Proceedings and Minutes of Evidence).

Canada. 1902. "Report on the Fourth Census of Canada, 1901." Ottawa.

Canada. 1902. "Report of the Royal Commission on Chinese and Japanese Immigration." *Sessional Papers*, no. 54, 1902.

Canada. 1905. "Fourth Census of Canada 1901." vol. 3: Manufactures. Ottawa: Dawson, Printer to the King's Most Excellent Majesty.

UNITED STATES

Cobb, John N. 1930. *Pacific Salmon Fisheries*. Bureau of Fisheries. Document No. 1092. Appendix XIII to the Report of the Commission of Fisheries for 1930. Washington: United States Government Printing Office. Fourth Edition.

Drucker, Philip. 1958. *The Native Brotherhoods: Modern Intertribal Organizations on the Northwest Coast*. Smithsonian Institution. Bureau of American Ethnology. Bulletin 168. Washington: United States Printing Office.

Rounsefell, George A., and George B. Kelez. 1938. *The Salmon and Salmon Fisheries of Swiftsure Bank, Puget Sound, and the Fraser River.* Bulletin No. 27. U.S. Department of Commerce. Bureau of Fisheries. From Bulletin of the Bureau of Fisheries. Volume XLIX. Washington: United States Government Printing Office.

SECONDARY SOURCES

Abrams, Philip. 1982. *Historical Sociology.* London: Open Books.

Adachi, Ken. 1976. *The Enemy That Never Was*. Toronto: McClelland & Stewart.

Alexander, Sally. 1979. "Women's Work in Nineteenth-Century London: A Study of the Years 1820–50." In *The Rights and Wrongs of Women*, ed. Juliet Mitchell and Ann Oakley, 59–111. England: Penguin.

Althusser, Louis. 1971. *Lenin and Philosophy and Other Essays*, trans. Ben Brewster. London: NLB.

Anderson, David. 1990. "Todd, Jacob Hunter." *Dictionary of Canadian Biography*, vol. 12, 1062–4, Francess Halpenny, Gen. Ed. Toronto: University of Toronto Press, vol. XII: 1062–4.

Arendt, Hannah. 1987. *Between Past and Present*. England: Penguin.

– 1972. *Crises of the Republic*. San Diego: Harcourt Brace Jovanovich.

– 1958. *The Human Condition*. Chicago: University of Chicago Press.

Armstrong, Pat, and Hugh Armstrong. 1983. "Beyond Sexless Class and Classless Sex: Towards Feminist Marxism." *Studies in Political Economy* 10(Winter):7–43.

– 1975. *The Double Ghetto: Canadian Women and Their Segregated Work*. Toronto: McClelland & Stewart.

Averkieva, Julia. 1971. "The Tlingit Indians." In *North American Indians in Historical Perspective*, ed. Eleanor Burke Leacock and Nancy Oestreich Lurie, 317–42. New York: Random House.

Barman, Jean. 1986. "Ethnicity in the Pursuit of Status: British Middle and Upper-Class Emigration to British Columbia in the Late Nineteenth and Early Twentieth Centuries." *Canadian Ethnic Studies* 18(1):32–51.

– 1991. *The West beyond the West: A History of British Columbia.* Toronto: University of Toronto Press.

Barrett, Michèle. 1985. *Women's Oppression Today: Problems in Marxist Feminist Analysis.* 2nd ed. London: Verso Editions and NLB.

Barrett, Stanley R. 1987. *Is God a Racist? The Right Wing in Canada.* Toronto: University of Toronto Press.

Barstow, Anne Llewellyn. 1994. *Witchcraze: A New History of the European Witch Hunts.* London: Pandora.

Baxandall, Rosalyn, Elizabeth Ewen, and Linda Gordon. 1976. "The Working Class Has Two Sexes." *Monthly Review* 28(3):1–9.

B.C. Packers, 1937–49. *Annual Reports.* British Columbia.

de Beauvoir, Simone. 1989. *The Second Sex.* New York: Vintage.

Berger, Thomas R. 1982. *Fragile Freedoms: Human Rights and Dissent in Canada.* Toronto: Clark, Irwin.

Bhaskar, Roy. 1983. "Dialectics." *A Dictionary of Marxist Thought*, edited by Tom Bottomore. Cambridge: Harvard University Press: 122–9.

Blyth, Gladys Young. 1991. *Salmon Canneries: British Columbia North Coast.* British Columbia: oolichan books.

Bolaria, B. Singh, and Peter S. Li. 1985. *Racial Oppression in Canada.* Toronto: Garamond Press.

Bonacich, Edna. 1976. "Advanced Capitalism and Black/White Race Relations in the United States: A Split Labor Market Interpretation." *American Sociological Review* 41:34–51.

– 1980. "Class Approaches to Ethnicity and Race." *The Insurgent Sociologist* 10(2):9–23.

– 1979. "The Past, Present, and Future of Split Labor Market Theory." *Research in Race and Ethnic Relations* 1:17–64.

Boswell, Terry E. 1986. "A Split Labor Market Analysis of Discrimination against Chinese Immigrants, 1850–1882." *American Sociological Review* 51(3):352–71.

Braverman, Harry. 1974. *Labor and Monopoly Capital: The Degradation of Work in the Twentieth Century.* New York: Monthly Review Press.

Bridges, Amy Beth. 1974. "Nicos Poulantzas and the Marxist Theory of the State." *Politics and Society* 4(2):161–90.

Burawoy, Michael. 1978. "Toward a Marxist Theory of the Labor Process: Braverman and Beyond." *Politics and Society* 8(2):267–312.

Burns, Janet. 1985. "Work Organization and the Derivation of Culture: Commercial Fishermen in Greater Vancouver." Paper presented at annual meetings Canadian Sociology and Anthropology Association. Montreal: Université de Montréal.

Carstens, Peter. 1991. *The Queen's People: A Study of Hegemony, Coercion, and Accommodation among the Okanagan of Canada*. Toronto: University of Toronto Press.

Cernetig, Miro. 1994a. "Fishery controls left gaping holes." *Globe and Mail*, Toronto, 20 Sept., A1, A4.

– 1994b. "Missing Pacific salmon put Ottawa on hook." *Globe and Mail*, Toronto, 16 Sept., A1, A8.

Chan, Anthony B. 1983. *Gold Mountain: The Chinese in the New World*. Vancouver: New Star Books.

Charles, Lindsey, and Lorna Duffin, eds. 1985. *Women and Work in Pre-Industrial England*. London: Croom Helm.

Chevillard, Nicole, and Sébastien Leconte. 1986. "The Dawn of Lineage Societies: The Origins of Women's Oppression." In *Women's Work, Men's Property: The Origins of Gender and Class*, ed. Stephanie Coontz and Peta Henderson, 76–107. London: Verso.

Clement, Wallace. 1986. *The Struggle to Organize: Resistance in Canada's Fishery*. Toronto: McClelland & Stewart.

Connelly, Pat. 1978. *Last Hired, First Fired: Women and the Canadian Labour Force*. Toronto: Women's Educational Press.

Connelly, M. Patricia, and Martha MacDonald. 1985. "Women and Development: The More Things Change the More they Stay the Same." Paper presented to the Conference on Women and Offshore Oil. St. John's Memorial University of Newfoundland.

– 1983. "Women's Work: Domestic and Wage Labour in a Nova Scotia Community." *Studies in Political Economy* 10(Winter):45–72.

Coontz, Stephanie, and Peta Henderson, eds. 1986. *Women's Work, Men's Property: The Origins of Gender and Class*. London: Verso.

Davies, Charles. 1987. *Bread Men: How the Westons Built an International Empire*. Toronto: Key Porter.

Dobb, Maurice. 1978. *Studies in the Development of Capitalism*. London: Routledge and Kegan Paul.

Dossa, Shiraz. 1989. *The Public Realm and the Public Self: The Political Theory of Hannah Arendt*. Waterloo: Wilfred Laurier University Press.

Drucker, Philip. 1963. *Indians of the Northwest Coast*. New York: Natural History Press.

Duff, Wilson. 1977. *The Indian History of British Columbia*, vol. 1. The Impact of the White Man. Victoria: British Columbia Provincial Museum.

Edwards, Richard. 1979. *Contested Terrain: The Transformation of the Workplace in the Twentieth Century*. New York: Basic Books.

Eichler, Margrit. 1988. *Families in Canada Today: Recent Changes and their Policy Consequences*, 2nd ed. Toronto: Gage.

Eisenstein, Zillah R., ed. 1979. *Capitalist Patriarchy and the Case for Socialist Feminism*. New York: Monthly Review Press.

Engels, Frederick. 1977. *The Condition of the Working-Class in England.* Moscow: Progress Publishers.

– 1981. *The Origin of the Family, Private Property and the State*, ed. and intro. Eleanor Burke Leacock. New York: International Publishers.

Evans, Hubert. 1954. *Mist on the River.* Toronto: McClelland & Stewart.

Farley, A.L. 1979. *Atlas of British Columbia: People, Environment, and Resource Use.* Vancouver: University of British Columbia Press.

Finkel, Alvin. 1977. "Origins of the Welfare State in Canada." In *The Canadian State: Political Economy and Political Power*, ed. Leo Panitch, 344–70. Toronto: University of Toronto Press.

Fisher, Robin. 1980. *Contact and Conflict: Indian-European Relations in British Columbia, 1774–1890.* Vancouver: University of British Columbia Press.

The Fisherman, 1937–80. Vancouver: The United Fishermen and Allied Workers Union.

Foley, Duncan. 1983. "Labor Power." *A Dictionary of Marxist Thought*, 265–7, ed. Tom Bottomore. Cambridge: Harvard University Press.

Fowke, Vernon. 1957. *The National Policy and the Wheat Economy.* Toronto: University of Toronto Press.

Freire, Paulo. 1990. *Pedagogy of the Oppressed.* New York: Continuum.

– 1985. *The Politics of Education: Culture, Power and Liberation.* Massachusetts: Bergin & Garvey.

French, Marilyn. 1985. *Beyond Power: On Women, Men and Morals.* New York: Summit.

Freud, Sigmund. 1982. *Civilization and its Discontents.* London: Hogarth Press.

Garcia, Wilma. 1984. *Mothers and Others: Myths of the Female in the Works of Melville, Twain and Hemingway.* New York: Peter Lang.

Garrod, Steve. 1984. "The Invisible People: Salmon Cannery Workers 1900–1935." Unpublished paper. Vancouver: Department of Anthropology and Sociology, University of British Columbia.

Gay, Peter. 1988. *Freud: A Life for our Time.* New York: Norton.

Giddens, Anthony. 1971. *Capitalism and Modern Social Theory.* Cambridge: Cambridge University Press.

– 1983a. *Central Problems in Social Theory: Action, Structure and Contradiction in Social Analysis.* Berkeley: University of California Press.

– 1986. *The Consitution of Society: Outline of the Theory of Structuration.* Berkeley: University of California Press.

– 1983b. *A Contemporary Critique of Historical Materialism*, vol. 1: Power, Property and the State. Berkeley: University of California Press.

– 1976. *New Rules of Sociological Method.* London: Hutchinson.

– 1977. *Studies in Social and Political Theory.* London: Hutchinson.

Gladstein, Mimi Reisel. 1986. *The Indestrutible Woman in Faulkner, Hemingway, and Steinbeck.* Michigan: U.M.I.

Gladstone, Percy Henry. 1959. "Industrial Disputes in the Commercial Fisheries of British Columbia," M.A. thesis, University of British Columbia.

Gold, David A., Clarence Y.H. Lo, and Eric Olin Wright. 1975. "Recent Developments in Marxist Theories of the Capitalist State." *Monthly Review* 27(5):29–43; 27(6):36–51.

Gough, Ian. 1979. *The Political Economy of the Welfare State*. London: Macmillan.

Gough, Kathleen. 1975. "The Origin of the Family." In *Toward an Anthropology of Women*, ed. Rayna Reiter, 51–76. New York: Monthly Review Press.

Gregory, H., and K. Barnes. 1939. *North Pacific Fisheries*. San Francisco: American Council of the Institute of Pacific Relations.

Grejda, Edward S. 1974. *The Common Continent of Men: Racial Equality in the Writings of Herman Melville*. New York: National University Publications, Kennicat Press.

Guppy, Neil. "Labouring on Shore: Transforming Uncommon Property into Marketable Products." In *Uncommon Property: The Fishing and Fish-Processing Industries in British Columbia*, ed. Patricia Marchak, Neil Guppy, and John McMullan, 199–222. Toronto: Methuen.

Hale, Sylvia M. 1990. *Controversies in Sociology: A Canadian Introduction*. Mississauga: Copp Clark Pitman.

Hamilton, Roberta, and Michèle Barrett, eds. 1986. *The Politics of Diversity: Feminism, Marxism and Nationalism*. Montreal: Book Center.

Held, David. 1989. *Political Theory and the Modern State: Essays on State, Power, and Democracy*. California: Stanford University Press.

Himmelweit, Susan. 1983. "Mode of Production." *A Dictionary of Marxist Thought*, 335–7, ed. Tom Bottomore. Cambridge: Harvard University Press.

Hobsbawm, E.J. 1981. *Industry and Empire: From 1750 to the Present Day*. England: Penguin.

– 1980. "The Revival of Narrative: Some Comments." *Past and Present* 86:3–8.

Iino, Masako. 1984. "Japan's Reaction to the Vancouver Riot of 1907." *BC Studies* 60:28–47.

Illich, Ivan. 1982. *Gender*. New York: Pantheon.

Karcher, Carolyn L. 1980. *Shadow over the Promised Land: Slavery, Race, and Violence in Melville's America*. Baton Rouge: Louisiana State University Press.

Kessler, Suzanne J., and Wendy McKenna. 1985. *Gender: An Ethnomethodological Approach*. Chicago: The University of Chicago Press.

Knight, Rolf. 1978. *Indians at Work: An Informal History of Native Indian Labour in British Columbia 1858–1930*. Vancouver: New Star Books.

– and Maya Koizumi. 1976. *A Man of Our Times: The Life-History of a Japanese-Canadian Fisherman*. Vancouver: New Star Books.

Knuttila, Murray. 1987. *State Theories: From Liberalism to the Challenge of Feminism*. Toronto: Garamond.

Kobayashi, Audrey, 1984. "Emigration to Canada and Development of the Residential Landscape in a Japanese Village: The Paradox of the Sojourner." *Canadian Ethnic Studies* 16(3):111–31.

Laclau, Ernesto. 1975. "The Specificity of the Political: The Poulantzas-Miliband Debate." *Economy and Society* 4:87–110.

Ladner, Edna G. 1980. *Mosaic Fragments: From the Memoirs of T. Ellis Ladner (1871–1958)*. NP: author.

Ladner, T. Ellis. 1979. *Above the Sand Heads: A vivid account of life on the Delta of the Fraser River, 1868–1900*. British Columbia: Friesen.

Lamb, W. Kaye. 1982. "Anderson, Alexander Caulfield." *Dictionary of Canadian Biography*, vol. 11, 16–18, Francess G. Halpenny, Gen. Ed. Toronto, University of Toronto Press.

Laslett, Peter. 1976. "The Wrong Way through the Telescope: A Note on Literary Evidence in Sociology and in Historical Sociology." *British Journal of Sociology* 27(3):319–42.

Leacock, Eleanor Burke. 1954. *The Montaignais 'Hunting Territory' and the Fur Trade*. Washington, DC: American Anthropological Association, Memoir 78.

– 1981. *Myths of Male Dominance: Collected Articles on Women Cross-Culturally*. New York: Monthly Review Press.

Leibowitz, Lila. 1986. "In the Beginning ...: The Origins of the Sexual Division of Labour and the Development of the First Human Societies." In *Women's Work, Men's Property: The Origins of Gender and Class*, ed. Stephanie Coontz and Peta Henderson, 43–75. London: Verso.

Lerner, Gerda. 1986. *The Creation of Patriarchy*. New York: Oxford University Press.

Lewis, Jane. 1984. *Women and England 1870–1950: Sexual Divisions and Social Change*. Sussex: Wheatsheaf Books.

Li, Peter S. 1979. "A Historical Approach to Ethnic Stratification: The Case of the Chinese in Canada, 1858–1930." *Canadian Review of Sociology and Anthropology* 16(3):320–32.

Luxton, Meg. 1980. *More than a Labour of Love: Three Generations of Women's Work in the Home*. Toronto: Women's Press.

– 1986. "Two Hands for the Clock: Changing Patterns in the Gendered Division of Labour in the Home." *The Politics of Diversity: Feminism, Marxism and Nationalism*, ed. Roberta Hamilton and Michèle Barrett, 35–52. Montreal: Book Center.

Lyons, Cicely. 1969. *Salmon: Our Heritage*. Vancouver: Mitchell Press.

McCunn, Ruthanne Lum. 1988. *Thousand Pieces of Gold*. Boston: Beacon.

MacDonald, Martha, and M. Patricia Connelly. 1985. "A Leaner Meaner Industry: A Case Study of 'Restructuring' in the Nova Scotia Fishery." Paper presented to the Workshop on the Political Implication of Industrial Restructuring: Primary Producers in Atlantic Canada. Kingston: Queen's University.

McDonald, Robert A.J. 1981. "Victoria, Vancouver and the Economic Development of British Columbia, 1886–1914." In *British Columbia: Historical Readings*, ed. W. Peter Ward and Robert A.J. McDonald, 369–95. Vancouver: Douglas & McIntyre.

MacDowell, Laurel Sefton. 1978. "The Formation of the Canadian Industrial Relations System During World War Two." *Labour/Le Travail* 3:175–96.

McFeat, Tom. 1978. *Indians of the North Pacific Coast*. Toronto: Macmillan.

McInnis, Craig. 1989. "B.C. strike threatens fish trade, packers say." *The Globe and Mail*, Toronto, 1 Aug., A1.

McKervill, Hugh W. 1967. *The Salmon People: The Story of Canada's West Coast Salmon Fishing Industry*. Sidney, BC: Gray's Publishing.

McLellan, David, ed. 1982. *Karl Marx: Selected Writings*. London: Oxford University Press.

McMullan, John. 1987. "State, Capital and the B.C. Salmon-Fishing Industry." In *Uncommon Property*, ed. M. Patricia Marchak, Neil Guppy, and John McMullan, 107–52. Toronto: Methuen.

Macpherson, C.B. 1978. *Property: Mainstream and Critical Positions*. Toronto: University of Toronto Press.

– 1979. *The Real World of Democracy*. Toronto: Canadian Broadcasting Corporation.

Ma, L. Eve Armentrout. 1983. "A Chinese Statesman in Canada, 1903: Translated from the Journal of Liang Ch'i-ch'as." *BC Studies* 59:28–43.

Marchak, M. Patricia. 1981. *Ideological Perspectives on Canada*, 2nd ed. Toronto: McGraw-Hill Ryerson.

– 1989. "What Happens When Common Property Becomes Uncommon?" *BC Studies* 80:3–23.

– Neil Guppy and John McMullan, eds. 1987. *Uncommon Property: The Fishing and Fish-Processing Industries in British Columbia*. Toronto: Methuen: 223–48; and 3–31.

Marlatt, Daphne, and Maya Koizumi, 1975. *Steveston Recollected: A Japanese-Canadian History*. British Columbia: Provincial Archives of British Columbia.

Marx, Karl. 1967. *Capital*. vol. 1, *A Critical Analysis of Capitalist Production*, ed. Frederick Engels. New York: International Publishers.

– 1988. *Economic and Philosophic Manuscripts of 1844*. Moscow: Progress Publishers.

– and Frederick Engels. 1988. *The German Ideology*, including "Introduction to a Critique of Political Economy," ed. and intro. C.J. Arthur. New York: International Publishers.

– 1973. *Grundrisse: Foundations of the Critique of Political Economy*, translated with a foreward by Martin Nicolaus. England: Penguin.

Mead, Margaret. 1963. *Sex and Temperament in Three Primitive Societies*. New York: Morrow Quill.

Meggs, Geoff. 1991. *Salmon: The Decline of the British Columbia Fishery.* Vancouver: Douglas & McIntyre.

Mies, Maria, and Vandana Shiva. 1993. *Ecofeminism.* London: Zed Books.

Miles, Robert. 1989. *Racism.* London: Routledge.

Miliband, Ralph. 1972. "Reply to Nicos Poulantzas." In *Ideology in Social Science: Readings in Critical Social Theory*, ed. Robin Blackburn, 253–62. Glasgow: Fontana.

– 1980. *The State in Capitalist Society: The Analysis of the Western System of Power.* London: Quartet Books.

Millerd, Frank W. 1988. "Windjammers to Eighteen Wheelers: The Impact of Changes in Transportation Technology on the Development of British Columbia's Fishing Industry." *BC Studies* 78:28–52.

Mills, Patricia Jagentowicz. 1987. *Woman, Nature, and Psyche.* New Haven: Yale University Press.

Mitchell, Alanna. 1992. "Keeping house worth billions." *Globe and Mail*, Toronto, 19 June, A1.

Mitchell, Juliet. 1975. *Psychoanalysis and Feminism: Freud, Reich, Laing and Women.* New York: Vintage Books.

Mohun, Simon. 1983. "Division of Labour," *A Dictionary of Marxist Thought*, 131–4. ed. Tom Bottomore. Cambridge: Harvard University Press.

– 1983. "Value." *A Dictionary of Marxist Thought*, 507–11, ed. Tom Bottomore. Cambridge: Harvard University Press.

Muszynski, Alicja. 1986a. "Class Formation and Class Consciousness: The Making of Shoreworkers in the B.C. Fishing Industry." *Studies in Political Economy* 20:85–116.

– 1986b. "The Creation and Organization of Cheap Wage Labour in the British Columbia Fishing Industry," PH.D. diss., Department of Sociology, University of British Columbia.

– 1987a. "Major Processors to 1940 and Early Labour Force: Historical Notes." In *Uncommon Property*, ed. M. Patricia Marchak, Neil Guppy, and John McMullan, 46–65. Toronto: Methuen.

– 1984. "The Organization of Women and Ethnic Minorities in a Resource Industry: A Case Study of the Unionization of Shoreworkers in the B.C. Fishing Industry 1937–1949." *Journal of Canadian Studies* 19(1):89–107.

– 1988. "Race and Gender: Structural Developments in the Formation of British Columbia's Salmon Cannery Labour Forces." *The Canadian Journal of Sociology* 13(2):103–20.

– 1987b. "Shoreworkers and UFAWU Organization: Struggles between Fishers and Plant Workers within the Union." In *Uncommon Property*, ed. M. Patricia Marchak, Neil Guppy, and John McMullan, 279–90. Toronto: Methuen.

– 1991. "What Is Patriarchy?" In *Race, Class, Gender: Bonds and Barriers*, ed. Jesse Vorst, Tanya Das Gupta, et al., 64–87. Toronto: Between the Lines.

Naylor, Tom. 1975. *The History of Canadian Business 1867–1914*, 2 volumes. Toronto: Lorimer.

Newell, Dianne, ed. 1989. *The Development of the Pacific Salmon-Canning Industry: A Grown Man's Game*. Montreal: McGill-Queen's University Press.

– 1993. *Tangled Webs of History: Indians and the Law in Canada's Pacific Coast Fisheries*. Toronto: University of Toronto Press.

North, George. 1974. *A Ripple, A Wave: The story of Union Organization in the B.C. Fishing Industry*, rev. and ed. Harold Griffin. Vancouver: Fisherman Publishing Society.

O'Brien, Mary. 1983. *The Politics of Reproduction*. Boston: Routledge and Kegan Paul.

Pacific Fisherman. 1952 (August). 50(9).

Palmer, Bryan D. 1983. *Working-Class Experience: The Rise and Reconstitution of Canadian Labour, 1800–1980*. Toronto: Butterworth.

Panitch, Leo. 1981. "Trade Unions and the Capitalist State." *New Left Review* 125:21–43.

Parkin, Frank. 1982. *Max Weber*. Chichester: Ellis Horwood.

Pentland, H. Clare. 1981. *Labour and Capital in Canada 1650–1860*. Toronto: Lorimer.

Piddocke, Stuart. 1969. "The Potlatch System of the Southern Kwakiutl: A New Perspective." In *Environment and Cultural Behavior: Ecological Studies in Cultural Anthropology*, ed. Andrew P. Vayda, 130–56. Garden City: Natural History Press.

Pinkerton, Evelyn. 1987a. "Competition Among B.C. Fish-Processing Firms." In *Uncommon Property*, ed. M. Patricia Marchak, Neil Guppy, and John McMullan, 66–91. Toronto: Methuen.

– 1987b. "Indians in the Fishing Industry." In *Uncommon Property*, ed. M. Patricia Marchak, Neil Guppy, and John McMullan, 249–69. Toronto: Methuen.

Piore, Michael J. 1979. *Birds of Passage: Migrant Labor and Industrial Societies*. Cambridge: Cambridge University Press.

Poulantzas, Nicos. 1976. "The Capitalist State: A Reply to Miliband." *New Left Review* 95:63–83.

– 1978. *Political Power and Social Classes*. London: New Left Books.

– 1972. "The Problem of the Capitalist State." *Ideology in Social Science: Reading in Critical Social Theory*, ed. Robin Blackburn, 238–53. Glasgow: Fontana.

Prentice, Alison, Paula Bourne, et al. 1988. *Canadian Women: A History*. Toronto: Harcourt Brace Jovanovich.

Ralston, Harry Keith. 1972. "Deas, John Sullivan." *Dictionary of Canadian Biography*, 216, ed. Marc La Terreur. Toronto: University of Toronto Press.

– 1977. "John Sullivan Deas: A Black Entrepreneur in British Columbia Salmon Canning." *B.C. Studies* 32:64–78.

– 1965. "The 1900 Strike of Fraser River Sockeye Salmon Fishermen," M.A. thesis, The University of British Columbia.

– 1981. "Patterns of Trade and Investment on the Pacific Coast, 1867–1892: The Case of the British Columbia Salmon Canning Industry." In *British Columbia: Historical Readings*, ed. Peter Ward and Robert A.J. McDonald, 269–305. Vancouver: Douglas & McIntyre.

Reid, David J. 1981. "Company Mergers in the Fraser River Salmon Canning Industry, 1885–1902." *British Columbia: Historical Readings*, ed. Ward and McDonald, 269–305. Vancouver: Douglas & McIntyre.

– 1973. "The Development of the Fraser River Salmon Canning Industry, 1885 to 1913." Vancouver: Department of the Environment NOS/ECON 4–73.

Reiter, Ester. 1991. *Making Fast Food: From the Frying Pan into the Fryer.* Montreal: McGill-Queen's University Press.

Reiter, Rayna R., ed. 1975. *Toward an Anthropology of Women*. New York: Monthly Review Press.

Rowbotham, Sheila. 1976. *Hidden from History: Rediscovering Women in History from the 17th Century to the Present*. New York: Vintage Books.

Roy, Patricia E. 1976. "The Oriental 'Menace' in British Columbia." In *Historical Essays on British Columbia*, ed. J. Friesen and H.K. Ralston, 243–55. Toronto: McClelland & Stewart.

– 1989. *A White Man's Province: British Columbia Politicians and Chinese and Japanese Immigrants, 1885–1914*. Vancouver: University of British Columbia Press.

Rubin, Gayle. 1975. "The Traffic in Women: Notes on the 'Political Economy' of Sex." In *Toward an Anthropology of Women*, ed. Rayna Reiter, 157–210. New York: Monthly Review Press.

Safa, Helen I. 1986. "Runaway Shops and Female Employment: The Search for Cheap Labor." In *Women's Work: Development and the Division of Labor by Gender*, ed. Eleanor Leacock, Helen I. Safa, and Contributors, 58–71. Bergin & Garvey Publishers.

Saliou, Monique. 1986. "The Process of Women's Subordination in Primitive and Archaic Greece." In *Women's Work, Men's Property: The Origins of Gender and Class*, ed. Stephanie Coontz and Peta Henderson, 169–206. London: Verso.

Sayers, Janet. 1982. *Biological Politics: Feminist and Anti-Feminist Perspectives*. London: Tavistock Publications.

– 1986. *Sexual Contradictions: Psychology, Psychoanalysis, and Feminism*. London: Tavistock Publications.

Schmidt, Alfred. 1973. *The Concept of Nature in Marx*. London: NLB.

Shaikh, Anwar. 1978. "An Introduction to the History of Crisis Theories." *U.S. Capitalism in Crisis*, 219–41. New York: Union for Radical Political Economics.

Skogan, Joan. 1983. *Skeena: A River Remembered*. Vancouver: British Columbia Packers.

Smith, Dorothy E. 1987. *The Everyday World as Problematic: A Feminist Sociology*. Toronto: University of Toronto Press.

Smith, Paul. 1980. "Domestic Labour and Marx's Theory of Value." In *Feminism and Materialism: Women and Modes of Production*, ed. Annette Kuhn and AnnMarie Wolpe, 198–219. London: Routledge and Kegan Paul.

Sparrow, Leona. 1976. "Work Histories of a Coast Salish Couple," M.A. thesis, University of British Columbia.

Stacey, Duncan A. 1982. *Sockeye and Tinplate: Technological Change in the Fraser River Canning Industry 1871–1912*. British Columbia: British Columbia Provincial Museum.

Stainsby, Gillian Mary. 1991. "'It's the Smell of Money': Women Shoreworkers of British Columbia," M.A. thesis, Simon Fraser University.

Stevenson, Garth. 1977. "Federalism and the Political Economy of the Canadian State." In *The Canadian State*, ed. Leo Panitch, 71–100. Toronto: University of Toronto Press.

Steltzer, Ulli, and Catherine Kerr. 1982. *Coast of Many Faces*. Vancouver: Douglas & McIntyre.

Suttles, Wayne. 1960. "Affinal Ties, Subsistence and Prestige among the Coast Salish." *American Anthropologist* 62:296–305.

– 1987. *Coast Salish Essays*. Seattle: University of Washington Press.

Tan, Jin, and Patricia E. Roy. 1985. *The Chinese in Canada*. Ottawa: Canadian Historical Association.

Tennant, Paul. 1990. *Aboriginal Peoples and Politics: The Indian Land Question in British Columbia, 1849–1989*. Vancouver: University of British Columbia Press.

Thomas, Robert J. 1982. "Citizenship and Gender in Work Organization: Some Considerations for Theories of the Labor Process." In *Marxist Inquiries: Studies of Labor, Class and States*, ed. Michael Burawoy and Theda Skocpol, s86–112. Chicago: University of Chicago Press.

Thompson, E.P. 1979. *The Making of the English Working Class*. Great Britain: Penguin.

– 1978. *The Poverty of Theory and Other Essays*. New York: Monthly Review Press.

United Fishermen and Allied Workers Union (UFAWU). 1984. "The Future of the B.C. Fishing Industry," vol. 2. Vancouver: UFAWU.

Ursel, Jane. 1992. *Private Lives, Public Policy: 100 Years of State Intervention in the Family*. Toronto: Women's Press.

Waldhorn, Arthur. 1972. *A Reader's Guide to Ernest Hemingway*. New York: Farrar, Straus & Giroux.

Ward, W. Peter. 1978. *White Canada Forever: Popular Attitudes and Public Policy Toward Orientals in British Columbia.* Montreal: McGill-Queen's University Press.

– 1982. *The Japanese in Canada.* Ottawa: Canadian Historical Association.

Warriner, Keith. 1987. "Regionalism, Dependence, and the B.C. Fisheries: Historical Development and Recent Trends." In *Uncommon Property,* ed. M. Patricia Marchak, Neil Guppy, and John McMullan, 326–49. Toronto: Methuen.

Weinbaum, Batya, and Amy Bridges. 1976. "The Other Side of the Paycheck: Monopoly Capital and the Structure of Consumption." *Monthly Review* 28(3):88–103.

Wickberg, Edgar, ed. 1982. *From China to Canada: A History of the Chinese Communities in Canada.* Toronto: McClelland & Stewart.

Williams, Raymond. 1981. *Keywords: A Vocabulary of Culture and Society.* Great Britain: Fontana/Croom Helm.

– 1978. *Marxism and Literature.* Oxford: Oxford University Press.

Wolf, Eric. 1982. *Europe and the People Without History.* Berkeley: University of California Press.

Wright, Erik Olin. 1979. *Class, Crisis and the State.* London: Verso.

Yee, Paul. 1984. "Business Devices from Two Worlds: The Chinese in Early Vancouver." *BC Studies* 62:44–67.

– 1986. "Sam Kee: A Chinese Business in Early Vancouver." *BC Studies* 69/70:70–96.

Yung, Judy. 1987. *Chinese Women of America: A Pictorial History.* Seattle: University of Washington Press.

Zaretsky, Eli. 1976. *Capitalism, the Family, and Personal Life.* New York: Harper and Row.

Index